Contents

Letts

EDUCATIONAL

ADVANCED SUBSIDIARY

Revise AS

Geography

Author

Peter Goddard

Specification lists

AQA A Geography

UNIT	SPECIFICATION TOPIC	CHAPTER REFERENCE	STUDIED IN CLASS	REVISED	PRACTICE QUESTIONS
Unit 1 Core concepts in physical geography	**Water on the land**				
	Systems and river regimes	1.1			
	Channel processes and landforms	1.2			
	Flooding as a hazard	1.3			
	River basin management	1.3			
	Climatic hazards and change				
	Cost and benefits of weather and climate	3.2			
	Climatic hazards	3.2			
	Climatic change	3.2			
	Energy and life				
	Systems, flows and cycles	5.2			
	Succession and climax	5.2			
	Soils	5.1			
	Human activity and soils	5.1			
Unit 2 Core concepts in human geography	**Population dynamics**				
	Population dynamics	7.1, 7.2, 7.3, 7.4			
	Distribution and density	7.1			
	Population change	7.1			
	Migration	7.3			
	Population structure	7.2			
	Settlement processes and pattern				
	Urbanisation and sub-urbanisation	6.3			
	Counter and re-urbanisation	6.3			
	Settlement structure	6.2			
	Size and spacing of settlements	6.1			
	Economic activity				
	Components and primary activity	8.1, 8.3			
	Secondary activity	8.1, 8.3			
	Tertiary activity	8.1			
	Quartenary and quinary activity	8.1			
Unit 3	See below				

Examination analysis

Module 1 Exam Three structured physical stimulus response questions and one extended prose question.		1 hr 30 min test	35%
Module 2 Exam Three structured human stimulus response questions and one extended prose question.		1 hr 30 min test	35%
Module 3 Skills paper Subject content from Modules 1 and 2		1 hr 15 min test	30%

AQA B Geography

UNIT	SPECIFICATION TOPIC	CHAPTER REFERENCE	STUDIED IN CLASS	REVISED	PRACTICE QUESTIONS
Unit A The dynamics of change	***Physical geography – short-term and local change***				
	Atmospheric, geomorphological and human processes affecting the drainage basin	1.1, 1.2			
	People and the environment – population and resources				
	The environmental and social issues relating to population and resources	6.3, 7.4			
	The environmental and social issues relating to energy resources in LEDCs and MEDCs	8.1			
	Human geography – change in the UK in the last 30 years				
	Changing sectoral and spatial organisation of business	8.2			
	Changes in the social and demographic environments within urban areas	6.3			
Unit B The physical options – study one of options P, Q or R	**P Glacial environments**				
	The effects of glacial, fluvio-glacial and periglacial activity				
	Q Coastal environments				
	The interaction of marine and sub-aerial processes and distinctive landforms	2.1			
	R Urban physical environments				
	Weather and climate in urban areas and surrounds	3.2			
	The ecology of urban areas and attitudes to conservation in urban areas	6.3			
Unit C The human options – study one of S or T	**S Urban change in the UK and wider world in the last 30 years**				
	Economic and social reasons for population movement in MEDCs and LEDCs the effects of such movements	7.3			
	Inner city decline and regeneration in the UK	6.3			
	T The historical rural and urban landscapes of England and Wales				
	The evolution of rural landscapes	6.1			
	The evolution of urban landscapes	6.1			
	Urban and rural landscapes as resources	6.1, 8.1			

Examination analysis

Module 1 Exam Three out of three structured questions to complete.   1 hr 30 min test   40%

Module 2 Exam One long structured question on one option from a choice of three.   1 hr 15 min test   30%

Module 3 Exam One long structured question on one option from a choice of two.   1 hr 15 min test   30%

OCR A Geography

UNIT	SPECIFICATION TOPIC	CHAPTER REFERENCE	STUDIED IN CLASS	REVISED	PRACTICE QUESTIONS
	Hydrological systems				
	Drainage basins features and processes	1.1			
	Hydrographs	1.3			
	Ecosystems				
	Components of the ecosystem	5.2			
	Case studies of succession	5.2			
Unit 1	**Atmospheric systems**				
	The Earth's atmosphere energy budget	3.1			
	Local energy budget	3.1			
	Lithosphere				
	Tectonics	4.1			
	Weathering	4.2			
	Slopes	4.2			
	Population				
	Distribution. Change in time and space	7			
Unit 2	**Settlement**				
	Pattern	6.1–6.3			
	Process	6.1–6.3			
	Change	6.3			
Unit 3	See below				

Examination analysis

Module 1	**Exam** Four sections in this module all studied and compulsorily tested using structured questions. There will be at least one question that will require more extended writing.	1 hr 30 min test 40%
Module 2	**Exam** Two compulsory areas of study using structured questions. There will be at least one question that will require more extended writing.	1 hr 15 min test 30%
Module 3	**Investigation/exam** Geographical investigation divided into two parts. A – Tests research and planning skills (a compulsory question is set) B – Presentation and analysis is assessed (choice of questions, two to do)	1 hr 15 min test 30%

OCR B Geography

UNIT	SPECIFICATION TOPIC	CHAPTER REFERENCE	STUDIED IN CLASS	REVISED	PRACTICE QUESTIONS
	Atmosphere and people				
	Energy	3.1			
	Moisture	3.1			
	The UK's weather	3.2			
	Affects on human activity	3.2			
Unit 1	**Landform systems and people**				
Physical	Weathering and slopes	4.2			
systems and	Fluvial processes and landforms	1.2			
their	Drainage basins are modified by man	1.3			
management	Management of the drainage basin	1.3			
	Coastal systems and people				
	Processes and landforms	2.1			
	Management and man	2.2			
	Dunes	5.2			
	Economic activity and change				
	Economic change over time	8.1, 8.2			
	Government policies	8.1			
	Benefits, problems and response	8.1			
Unit 2	**Settlement dynamics**				
	Size and situation/patterns	6.1, 6.2			
Human	Function	6.1			
systems and	Change	6.1, 6.3			
their	**Population and development**				
management	Population dynamics	7.1, 7.2			
	Demographic change	7.1			
	Management of resources	7.1, 7.4			
	Management of population change and economic development	7.4			
Unit 3	See below				

Examination analysis

Module 1 **Exam** Two sections to the paper. Section A – Three compulsory structured questions. Section B – Consists of concise essay questions. One out of two to be completed.   *1 hr 15 min test 33.3%*

Module 2 **Exam** Two sections to the paper. Part A – Three compulsory structured questions. Section B – Consists of concise essay questions. One out of two to be completed.   *1 hr 15 min test 33.3%*

Module 3 **Investigation/exam** Paper has two sections. Section A – Planning, implementing and evaluation. One concise essay based on, and referring to, a previously written 1000-word report. Section B – Collecting, manipulating and analysing. Two data-handling structured questions are set.   *1 hr 15 min test 33.3%*

Edexcel A Geography

UNIT	SPECIFICATION TOPIC	CHAPTER REFERENCE	STUDIED IN CLASS	REVISED	PRACTICE QUESTIONS
Unit 1 Physical environments	**Earth systems**				
	Plate tectonics	4.1			
	Igneous activity and the weathering of rocks	4.2			
	Fluvial environments				
	Hydrological cycle	1.1			
	Long- and short-term variations in discharge	1.3			
	Channel processes and landforms	1.2			
	Coastal environments				
	Coastal systems and processes	2.1			
	Coastal landforms	2.1			
	Coastal ecosystems	5.2			
Unit 2 Human environments	**Population characteristics**				
	Population distribution, structure and growth	7.1, 7.2			
	Settlement patterns				
	Site and situation	6.1			
	Patterns and hierarchies	6.1			
	Urban land-use patterns	6.1, 6.2			
	Population movements				
	Causes, consequences and effects of migration	7.3			
Unit 3	See below				

Examination analysis

Module 1	**Exam** Short structured questions, using data stimulus. Six questions are set, two on each area of study. Three questions are answered, one from each pair.	*1 hr 15 min test* *30%*
Module 2	**Exam** Short structured questions, using data stimulus. Six questions are set, two on each area of study. Three questions are answered, one from each pair.	*1 hr 15 min test* *30%*
Module 3 OR	**Coursework/exam A** A piece of individual coursework of 2500 to 3000 words in length. The title and plan to be approved by Edexcel. **B** – A written examination divided into two parts: (a) A compulsory structured question based on resource material. (b) A compulsory question relating to the candidate's own fieldwork experience.	*1 hr 15 min test* *40%*

Edexcel B Geography

UNIT	SPECIFICATION TOPIC	CHAPTER REFERENCE	STUDIED IN CLASS	REVISED	PRACTICE QUESTIONS
Unit 1 Changing landforms and their management	*Riverine environments*				
	Channel and catchment processes and their management	1.1, 1.2, 1.3			
	Hydrosere successions				
	Challenge of management and impacts of human usage	1.3			
	Coastal environments				
	Short-term coastal changes	2.2			
	Longer term impact of sea level change	2.2			
	Coastal ecosystems and the challenge of managing them	2.2			
	Estuaries and deltas	1.2			
Unit 2 Managing change in human environments	**Processes in rural and urban areas and how they vary in space and time**				
	Issues facing these areas	6.1, 6.3			
	Rural–urban inter-relationships	6.1			
	What futures these environments?	6.1, 6.3			
Unit 3	See below				

Examination analysis

Module 1	**Exam** Short structured questions, based on data stimulus material. Choice of questions and some extended writing will be required.	*1 hr 30 min test*	*33.3%*
Module 2	**Exam** Short structured questions, based on data stimulus material. Choice of questions and some extended writing will be required.	*1 hr 30 min test*	*33.3%*
Module 3	**Investigation** Environmental, and issue based, individual investigation of 2500 words. Group proposals approved by Edexcel. Action plan required and has to be followed by individual students.		*33.3%*

WJEC Geography

UNIT	SPECIFICATION TOPIC	CHAPTER REFERENCE	STUDIED IN CLASS	REVISED	PRACTICE QUESTIONS
	Drainage basins – floodplains				
	Systems and hydrographs	1.1, 1.2			
	Processes and features	1.2			
	Flooding	1.3			
	The interaction of man with this environment	1.3			
Unit 1 Processes and issues in physical geography	**Global processes, earthquakes and volcanic hazards**				
	Global tectonic processes	4.1			
	Tectonic hazards	4.1			
	Impacts of tectonic activity	4.1			
	Responses to tectonic	4.1			
	Small ecosystems				
	Systems, flows and cycles	5.2			
	Succession and climax	5.2			
	Soils, one only to be studied	5.1			
	Global dynamic population				
	Population data and DTM	7.1, 7.2			
	Age/sex structure	7.2			
	Growth	7.1			
	Population and resources	7.1, 7.4			
Unit 2 Processes and issues in human geography	**Rural change and counter urbanisation**				
	Counter-urbanisation	6.2			
	Changes in rural settlement	6.1			
	Growth and decline in urban areas				
	Population movements	6.2			
	Deprivation and segregation	6.2			
	The rural urban fringe	6.1			
	Decision making in the city	6.2			
Unit 3	See below				

Examination analysis

Module 1	**Exam**	Choice of one from two data response questions and one extended prose question, from two, using stimulus material.	1 hr 15 min test 33.3%
Module 2	**Exam**	Choice of one from two data response questions and one extended prose question, from two, using stimulus material.	1 hr 15 min test 33.3%
Module 3	**Written skills paper**	Application of skills acquired in Units 1 and 2, appropriate to geographical investigation.	1 hr 30 min test 33.3%

NICCEA Geography

UNIT	SPECIFICATION TOPIC	CHAPTER REFERENCE	STUDIED IN CLASS	REVISED	PRACTICE QUESTIONS
Unit 1 *Themes in physical geography*	**Fluvial environments**				
	Processes and features	1.2			
	The interaction of man with this environment	1.3			
	Atmosphere				
	Heating, motion and moisture	3.1			
	Mid-latitude weather systems	3.2			
	Extreme weather	3.2			
	Ecosystems				
	Systems, flows and cycles	5.2			
	Succession and climax	5.2			
	Soils	5.1			
	Human activity and soils	5.1			
Unit 2 *Themes in human geography*	**Population**				
	Population data	7.1			
	Age/sex structure	7.2			
	Population and resources	7.1, 7.4			
	Settlement				
	Distribution	6.1			
	Rural settlement	6.1			
	Urban settlement	6.1, 6.3			
	Development				
	The nature and measure of development	A2			
	Processes of development	A2			
Unit 3	See below				

Examination analysis

Module 1	**Exam**	Three compulsory structured questions and each question will have at least one extended element.	1 hr 30 min test 35%
Module 2	**Exam**	Three compulsory structured questions and each question will have at least one extended element.	1 hr 30 min test 35%
Module 3	**Skills paper**	Techniques in Geography paper. Consists of two multi-part compulsory questions. Some collected data is brought to the exam for examination externally, this to is compulsory.	1 hr 15 min test 30%

AS/A2 Level Geography courses

AS and A2

From September 2000 all A Level qualifications will comprise three units of AS (Advanced Subsidiary) assessment and three units of A2 assessment. This offers Geography students the opportunity to complete a freestanding AS course and to complete their geographical education, or to develop ideas, themes and concepts further into a full A Level course via the much more demanding and challenging A2 course.

How will you be tested?

Assessment units

For AS Geography, you will be tested by three assessment units. For the full A Level in Geography, you will take a further three units. AS Geography forms 50% of the assessment weighting for the full A Level.

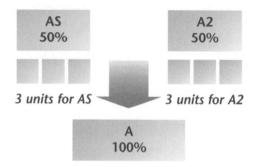

3 units for AS   3 units for A2

Each unit can normally be taken in either January or June. Alternatively, you can study the whole course before taking any of the unit tests. There is a lot of flexibility about when exams can be taken and the diagram below shows just some of the ways that the assessment units may be taken for AS and A Level Geography.

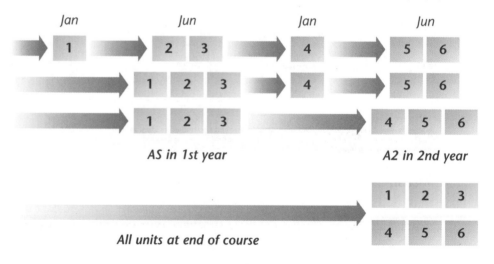

If you are disappointed with a module result, you can resit each module once. You will need to be very careful about when you take up a resit opportunity because you will have only one chance to improve your mark. The higher mark counts.

A2 and Synoptic assessment

After having studied AS Geography you may wish to continue studying Geography to A Level. For this you will need to take three further units of Geography at A2. Similar assessment arrangements apply except some units, those that draw together different parts of the course in a synoptic assessment, have to be assessed at the end of the course.

Coursework

· Fieldwork/coursework, in whatever guise, will form part of either AS or A2. There is no minimum weighting for coursework in Geography. Specifications vary between 20% and 40% over the two years of study.

Key skills

AS and A2 Geography specifications identify opportunities for developing and assessing key skills, where these are appropriate to the subject. The range of key skills is broad, incorporating Communication, ICT, Application of number, Improving own learning and performance, Working with others and Problem solving. However, the only key skill component that has to be assessed through Geography is Communication.

Should you choose to pursue the AS key skills qualification it would be possible, by following your AS Geography course, to reach and demonstrate Level 3 (the necessary level of achievement), in Communication, Application of number and ICT, and conceivably this could come from your normal class activities and by formal testing.

The Key Skill AS is a worthwhile qualification and demonstrates your ability to put your ideas across to other people, collect data and use up-to-date technology. The Key Skills AS, like all other ASs, is worth half the UCAS score of the more advanced A2 qualification.

Different types of questions in AS examinations

Short and extended prose/essay questions

A mainstay of the examination system, and one that makes a more prominent return in many of the new specifications. Essays in whatever guise can be the downfall for those who are unprepared.

The basic components of an AS/A2 Geography essay answer are as follows

- There is a correct and appropriate response to command words.
- There is relevant content to the question put forward.
- There is an approach which is both confident and direct.
- Work is paragraphed, with one correct, relevant and strong theme running through the essay.
- There is a structure to the essay, guided by an appropriate plan (written on the paper that is handed in).
- Case studies are used but not overused, and these draw on a variety of scales (local, regional and global, if required).
- Where a comment on process is needed comment is offered.
- Balance is achieved between too much fact and the more discursive aspects of the essay.
- Diagrams and sketches are used that move the essay on.
- With regard to structure, a basic introduction is important (it should capture the examiner's interest). An 'expansion' of ideas in the middle section of the essay (the 'meat' of the essay) should appear next, followed by a conclusion. (It is important this is not a repeat or a summary of what's gone previously.)
- That you have allowed enough time to sit back and correct mistakes and omissions.

Short and long structured questions (SLSQs)

All the examination boards are utilising these types of question in their assessment suites. Such questions are sub-divided up with each section building on the last.

With these questions the single most important thing to do is to read the rubric (the instructions for the exam). Answer the correct number of questions, and aim to spend time commensurate with those parts of the questions that carry most marks. Fill the spaces left for your answers, avoiding re-runs of the question. You have to quickly focus your thoughts and ideas, but don't hurry your actual answers! Perhaps small plans in the margin would help this process, certainly underline or highlight key parts of the question stem and command words.

Many SLSQs come with extra information, like maps, graphs and tables. These are included to help you, they are not just page decoration; detail from them is absolutely vital for your answers. Some of these 'additions' are covered in a little more detail below.

- **Photos** If these are offered, on the whole what tends to be set is a descriptive exercise; occasionally process-based questions are asked.
- **Diagrams** This is where a good understanding of the vocabulary of the

subject helps. Your labels, if more than a single word is required, must show understanding. The other way diagrams are used is of course for you to actually have to draw relevant and labelled offerings. You should have a reserve of simple AS standard diagrams stashed/stored away for use in both the structured and essay papers.

- **Graphs and tables** The data these offer is invariably going to have anomalies, differences, increases, decreases and periods of decline, trends and groupings. Questions that relate to such tables and graphs usually ask you to pick out and use relevant data in your responses.

- **Case studies** These are as important in structured papers as they are in essay papers. Clearly on this paper they need to be focused, short, relevant and to the point. Sources for your case studies vary, but obvious places to glean them from are newspapers (broad sheets are best!), the television and, if you are connected, the internet.

You should also be aware that SLSQs, if compared to essay writing, are completely different in their demands and the technique required to successfully answer them. Practice is important, to both bring on your technique and to ensure that you have the time issue, outlined above, completely sorted out. There are lots of opportunities to practice such questions in this book!

Ordnance Survey and SLSQs

Most boards use OS maps, though to a highly variable degree. Most seek to use them for analytical and interpretative purposes, usually as a 'tail' to SLSQs. In the past lots of map interpretation had been reduced to exercises in recognition and detection, a great shame as it is a core geographical skill. The AS utilises analysis and interpretation of the OS map to both extend and develop your geographical intellect.

Space prohibits the inclusion of OS extracts and exercises in this text, but what follows is an attempt to cover the approach and themes covered at AS Level.

On the whole maps are not chosen to spotlight 'classic' cultural landscapes, or to focus on specific areas or landforms. They are usually chosen to represent problem areas for which an interpretation can be offered on the basis of printed evidence from both the physical landscape and/or the rural and urban landscape. An exposure to foreign maps too, is just as important as OS work.

Typically questions ask for a description, analysis or synthesis of evidence presented by the OS map, or any combination of these three.

Descriptive accounts usually involve looking at physical or cultural elements on the map. The features are described in terms of their relief, vegetative cover, or settlement and communications might also be described.

Analysis in which map evidence is systematically analysed at an elementary level and in relation to other mapped distributions.

Synthesis in which analyses are collated to form a statement or interpretation of the physical and cultural features observed.

Written coursework investigation/examinations (WCI)

WCIs first appeared in the last revision of Advanced Level Geography. Its popularity with many centres has ensured a place for it with a number of the boards at AS (and A2).

Generally there is a selection of study area to be made (usually a human or physical choice). You will usually have access to pre-release material, which enables you to become familiar with the selected topic, the purpose of the study, the theory relevant to the study and some data related to the study and methods of data collection. It is also important to experience the topic firsthand through some fieldwork, and to apply this field information in the examination proper.

Question spotting should be avoided on this type of paper. It is only too easy to focus incorrectly on irrelevant sections; responses that are inappropriate to the set question gains few marks. It is also important to allocate time extremely carefully, writing to fulfil the requirements of the mark allocation, rather than concentrating on just one area. Incompleteness is to be avoided. The best responses are short and respond closely to the command words.

For many this type of approach works, but be warned it isn't an easy option, it entails as much work as the individual fieldwork study and is marked to the same standard!

In many specifications it is possible to take this route in Year 12 and to opt for a repeat or individual study in Year 13, if your first result is poor.

The investigative study/individual enquiry

Used in some form by all the boards, this personal piece of work requiring primary data collection is based on an issue, problem or question. An external moderator usually approves titles, and the finished work is of some 2500 to 4000 words depending on the specification you are following.

These investigations are a challenging task for you to undertake, but one which if small-scale and focused, local and accessible, and topical can be extremely satisfying and enjoyable.

Perhaps the toughest part of the individual study is the selection of the topic/title. It is advisable to look first at the specification, at past successful titles and importantly talk to your teachers. Then check the viability, scale and appropriateness of your ideas.

The typical route through the enquiry process is shown below:

You should read your board's specification for any special features that are required or needed by them in the individual study.

As the individual study is such an undertaking it is wise to plan and prepare a timetable over a longish period of time to ensure you bring the study to its proper conclusion. You'll need to be highly self-motivated and determined or you will under-perform.

A suggested timetable for a study handed in, in Year 13, runs thus:

Year 12	
September/November	proposal submitted
January/February	study title is finalised
March/April	background reading and research
May/July	thorough collection of data (two days minimum in the field)
August	processing of the data over the summer vacation
Year 13	
September	produce the draft report (based on your boards specifications)
October	add representative techniques and statistics (using a range of representative and manipulative techniques)
November	checking of report (parents or teachers)
December/January	final word processing and hand in (ensure you know how the board want it presented)

Do take advantage of the opportunity that the individual study offers: the chance to score highly in a part of the suite of assessments that you can control!

Exam technique

Links from GCSE

In order to study AS Geography it is of course desirable that you have studied Geography to GCSE Level. But it is emphatically not a requirement, though of course it helps! It is true that some topics covered at AS do develop ideas first encountered at GCSE, but many study areas are new and others develop ideas in new or different way. You will therefore not be at a disadvantage if you are new to the subject.

It is likely that your teachers will be looking for students that have an interest in the environment and the world around them, and want to follow a subject that is relevant to their own situation and life. Students who are willing to explore ideas in an enquiring and lively way, and who can pass on and communicate findings and ideas effectively, are most likely to get the most out of Geography, and this text!

What are examiners looking for?

All questions of whatever type seek to assess your appreciation and attitude to 'content', that is content, be it physical, human or regional in nature, of the specification you are involved with. There are certain common qualities that all boards look for in candidates.

1 A knowledge of facts, basic vocabulary, geographical concepts, processes and theories.

2 An ability to use information in an organised way, supported by appropriate case studies and examples.

3 The appreciation that all geography content is dynamic.

4 A range of skills understood and used in a range of geographical contexts.

5 An ability to comment and evaluate world issues and problems is paramount.

Quality of English is now integrated into the marking of papers. You must be clear and accurate in your use of English (i.e. spellings like desert, erosion and vegetation must be correct).

Some dos and don'ts

Examiners then, are trying to assess the degree to which you can demonstrate as many of the qualities listed above as possible. Common problems encountered by examiners when they are marking examination work include (and these are the areas that you need to avoid!):

• Candidates spending too much time on one question or a part of a question.

• Candidates who let words like coast or river trigger an 'all I know' type of response, avoiding the focus of the question.

• Candidates who over-learn favourite topics in the hope they will appear in the examination.

• Candidates who don't plan answers, whether they are in response to extended writing questions or structured questions. Quick simple plans are a must, rewriting the essay title on your answer sheet can also help.

• Candidates who fail to respond to command words. Don't describe if it says explain; more of this later.

- Candidates who attempt questions that lead to the snake pit: if you don't understand plastic deformation (glaciation), flocculation (soils) or fronto-genesis (atmosphere) avoid these questions!

- Candidates who don't respond to the vocabulary in the question, i.e. channels are different to valleys, glaciation is different to peri-glaciation, environmental hazards could mean pollution and/or biological and/or geomorphologic hazards, weathering is different to erosion, and so on...

- Candidates who have a limited reserve of case studies to draw upon. And those that choose case studies that obscure rather than illustrate points.

- Candidates who seem to be unaware of the ploys to reach the highest level or tier in a question.

- Candidates who fail to read the rubric of the examination, or fail to organise their time effectively enough over the whole examination. These candidates invariably show all the signs of panicking, i.e. questions unfinished and poor quality of language, uncorrected grammatical and spelling errors and major and avoidable omissions. It is important to re-read your completed examination!

- Candidates who have no real knowledge of their specification and how knowledge, understanding and skills fit each module/paper that they sit. Failure to practice past papers is also very obvious.

- Candidates who don't seek out and outline inter-relationships.

- Candidates who don't 'give' requested information, i.e. A named area within a city.

- Candidates who don't offer supporting diagrams and sketches.

What grade do you want?

The grade that is eventually awarded to you will depend upon the extent to which you have met the assessment objectives of the exam board you are studying with. Clearly, to gain the best possible mark you will have worked hard over the year. You will be highly determined and motivated. If you have identified weaknesses in your knowledge and understanding you will balance this by improving and building on your performance in other areas of the subject.

For a grade A

You will be a student who can:

- show a comprehensive knowledge of places, themes and environments
- understand how physical and human processes affect the above three areas
- show sound understanding of concepts, theories and principles
- understand a wide range of Geographical terminology
- understand how all of the above connect and be able to convey your understanding at a variety of scales.

For a grade C

You will be the student that produces sound, rather than competent answers. You may have some weakness in your understanding and knowledge and may be unsure of some terminology. You may synthesise and communicate your ideas and views less effectively than the A grade candidate.

- To improve you need to master and improve upon all of your weaknesses.
- You must prepare fully. Practice past questions.
- Hopefully, you will read around the subject a little more and keep up with current affairs.

For a grade E

You cannot afford to miss any marks! Even if you find the subject matter difficult to comprehend and would be content with an E, there are ways you can improve your prospects.

- Start by memorising the terminology of the subject.
- You must practice past questions. Being able to answer such questions even if it isn't in exam conditions is a great confidence-booster. On difficult questions try and answer the easier parts first, come back to the tougher parts later.
- It is likely that the areas of the subject that interest you the least are the areas where you experience the most difficulties, whether it be on a structured or essay paper. Such questions can be attempted and you will gain marks even if you only get part of the way through.

If you are working towards an A grade you need to keep at it and retain both your motivation and persistence. If you are not so fortunate, rest assured you will improve. Reading this text thoroughly, including the preliminary pages and the chapters themselves, and by completing the practice questions, will start this improvement process!

What marks do you need?

The table below shows how your average mark is translated.

average	80%	70%	60%	50%	40%
grade	A	B	C	D	E

Four steps to successful revision

Step 1: Understand

- Study the topic to be learned slowly. Make sure you understand the logic or important concepts.
- Mark up the text if necessary – underline, highlight and make notes.
- Re-read each paragraph slowly.

GO TO STEP 2
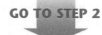

Step 2: Summarise

- Now make your own revision note summary:
  What is the main idea, theme or concept to be learned?
  What are the main points? How does the logic develop?
  Ask questions: Why? How? What next?
- Use bullet points, mind maps, patterned notes.
- Link ideas with mnemonics, mind maps, crazy stories.
- Note the title and date of the revision notes
  (e.g. Geography: Soils and ecosystems, 3rd March).
- Organise your notes carefully and keep them in a file.

This is now in **short-term memory**. You will forget 80% of it if you do not go to Step 3.
GO TO STEP 3, but first take a 10 minute break.

Step 3: Memorise

- Take 25 minute learning 'bites' with 5 minute breaks.
- After each 5 minute break test yourself:
  Cover the original revision note summary
  Write down the main points
  Speak out loud (record on tape)
  Tell someone else
  Repeat many times.

The material is well on its way to **long-term memory**.
You will forget 40% if you do not do step 4. **GO TO STEP 4**

Step 4: Track/Review

- Create a Revision Diary (one A4 page per day).
- Make a revision plan for the topic, e.g. 1 day later, 1 week later, 1 month later.
- Record your revision in your Revision Diary, e.g.
  Geography: Soils and ecosystems, 3rd March 25 minutes
  Geography: Soils and ecosystems, 5th March 15 minutes
  Geography: Soils and ecosystems, 3rd April 15 minutes
  ... and then at monthly intervals.

Water on the land

The following topics are covered in this chapter:

- *Rivers and the hydrological cycle*
- *River processes and landforms*
- *Water management*

1.1 Rivers and the hydrological cycle

After studying this section you should be able to understand:

- that water is continually entering and leaving the river system
- that this open system coexists within the closed hydrological system
- that drainage basins are more or less self-contained, and are therefore a convenient means of considering the effect and action of running water
- that most streams and rivers have quite simple origins

LEARNING SUMMARY

Rivers as open systems

AQA A	U1	EDEXCEL A	U1
AQA B	UA	EDEXCEL B	U1
OCR A	U1	WJEC	U1
OCR B	U1	NICCEA	U1

Rivers dominate the physical landscape, producing widespread changes. Rivers remove rock and material from mass wasting, particularly but not exclusively in the humid areas of the world. Rivers move under gravity in a channel transferring both water and sediment downstream. Rivers act as open systems. Stores and transfers vary in size according to changes in inputs of water. If all sections of the 'system' balance it is said to be in **dynamic equilibrium**

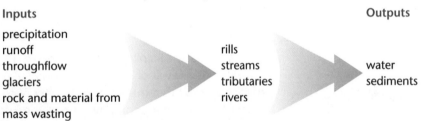

Inputs		Outputs
precipitation	rills	
runoff	streams	water
throughflow	tributaries	sediments
glaciers	rivers	
rock and material from mass wasting		

The hydrological cycle

What happens in the river is part of a cycle of events called the hydrological cycle (below).

A frequently asked area of study at AS.

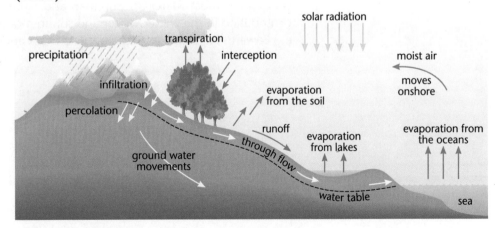

The river represents the flow of water overland, but two key 'hidden' aspects are of vital significance.

It is possible to compile graphs of precipitation, potential and actual evapotranspiration. These are called soil moisture budget graphs (below).

The soil moisture budget for Norfolk.

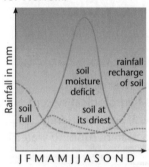

------ precipitation

············· actual evapotranspiration

——— potential evapotranspiration

A knowledge of the soil moisture budget and potential evapotranspiration allows irrigation 'need' to be calculated.

Infiltration

This is probably the most significant and fundamental portion of the drainage basin system, contributing as it does to groundwater storage, throughflow and stream channel recharge.

Infiltration rates are affected by the following.

- The **nature of precipitation**; its duration, total, the area covered, the size of raindrops, frequency and its chemical composition.

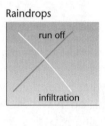
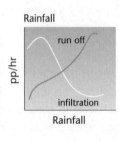
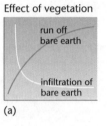
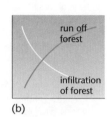

- The effect of **vegetation interception** depends on the nature of the vegetation. Pine forests, for example, take out as much as 96% of low-intensity rainfall, leaving only 4% to infiltrate.
- The effect of **depression storage**, the geological structure and man's agricultural activity. Water captured in the landscape, or plough-furrows infiltrates!
- **Evapotranspiration** – available heat, wind, the texture and depth of soils affect how much water returns to the atmosphere from exposed water surfaces and from vegetation.
- The **nature of the soils**, their permeability, porosity and texture.
- **Slope length and angle** which determine whether runoff or infiltration is the dominant process.

Groundwater

This appears and collects in permeable rocks, known as aquifers, which are saturated as a result of infiltration by rainfall. Water flows between the small grains of rock and soil that make up the aquifer. Variations in the shape of the water table reflect the surface topography; that is, the water table is near the surface in valleys and deep down in the hills. It is because of the topography of the landscape the groundwater 'flows'. The velocity of flow is proportional to the hydraulic gradient (simply the slope of the water table induced by the topography of the landscape); speeds are typically between 1m/yr to 1m/day. Most rivers gain their water/flow from a combination of surface runoff and groundwater discharge. Groundwater contributes most water in summer and autumn in the UK, surface runoff contributing in the wetter winter and spring months.

e.g Edwards Aquifer, San Antonio, USA, Chalk aquifers of Southern England.

1.2 River processes and landforms

After studying this section you should understand:

- *that effects of erosion, transport and deposition determine resultant landforms*

Fluvial processes

AQA A	U1	EDEXCEL A	U1
AQA B	UA	EDEXCEL B	U1
OCR A	U1	WJEC	U1
OCR B	U1	NICCEA	U1

River channels not only transfer water and sediment. The potential and kinetic energy within the water does the work of erosion, transportation and deposition within the river. Flow type affects the efficiency of a river to erode, transport or deposit. Two principle flow types are recognised, turbulent flow and laminar flow. Both are affected by friction.

Energy and flow

The ability of a river to perform its various feats of erosion, transportation and deposition is determined by the amount of energy it possesses. Water in rivers has both potential energy and kinetic energy.

Energy flow relationships

During periods of low flow, below bank stage flow or when rivers are at base flow, low energy conditions are experienced and little 'work' is done. Channel adjustments are most likely to occur during the rising phase of the storm hydrograph, when there is a lot of water in the system; it is at this point that erosion is at its height. Deposition occurs as discharge and energy declines on the falling limb of the storm hydrograph. A relationship exists between volume and velocity and individual particle erosion, transportation and deposition; this is displayed on the Hjulström Curve.

> *Increased energy/water input =* increased channel capacity and efficiency. Width and depth increase. Gradient increases. Velocity and throughput of water increases. Erosion and transportation at their most efficient.
>
> *Decreased energy/input of water subsides =* discharge and velocity decreases. Erosion ceases and transportation slows, deposition begins and the river becomes narrower and shallower.

KEY POINT

In summary, riverine landforms and landscapes go through long periods of stability, interspersed with short but rapid periods of change, related to variations in energy input and material surges.

The energy available to a river enables it to accomplish three main types of work:

- the land surface over which the river flows is both **eroded** and dissolved
- eroded and dissolved material is **transported** away
- the river **deposits** the material that has been carried.

There are strong links between these three components.

Erosion

To understand the important work that erosion completes in a river three areas have to be further explored, volume, velocity and load. The latter is of vital importance as a river 'charged' with sediment is able to wear the land surface away more effectively than one that is merely dissolving it away.

Process questions are frequently used at the start of structured questions.

Volume

Most streams and rivers obtain their water from rainfall or any of the other forms of precipitation. This precipitation either evaporates, soaks in or contributes to the runoff or drainage of the land surface. As rivers flow from high (source areas) to lower areas (the mouth, usually the sea, unless the river has entered an arid basin) their volume increases as contributions from other parts of the drainage basin via tributaries are added. There can be variations in a river's volume relating to seasonality of rainfall (in monsoon areas), the contribution of snow melt and of springs and groundwater.

$Q = AV$

(A = cross sectional area
V = velocity)

> **KEY POINT**
>
> *Discharge* (Q) is defined as the volume of water passing a particular point in a river in a unit of time, expressed as m^3/s^{-1}, or in cumecs.

Velocity

Velocity is more or less constant along the length of the river. It is true that steeper slopes do encourage higher velocities, but the larger channels of the lower course exert less friction than the small channels of the upper course, causing a relative increase in velocity and allowing the river to become much more efficient.

Load

Know this for AS.
Remember the Hjulström curve?

> **KEY POINT**
>
> Put simply:
> Volume of water carried + velocity of this water = energy of the river. Energy availability determines the capacity (total load) and calibre (the weight/size dimensions of individual particles).

Two points here:

- Small streams carry a greater quantity of fine material than of coarse. Conversely, large rivers with more available energy carry larger/coarser material.
- The erosive power of a river is for the most part determined by the 'charge' of debris it carries. Running water has restricted erosional ability. The proviso being that with increased load a river is more likely to begin to aggrade, or deposit.

The mechanism of erosion

For a particle to be used in the erosional process it has to be removed from the bed and banks of the river by erosion and **entrained**. This entrained material is acquired in three ways.

Erosion = wearing away + movement

Erosion + weathering = denudation

Weathering destroys rock *in situ*, little movement is involved.

- **Vertical erosion** deepens channels, aided by weathering mass movement and soil creep. Characteristics of channels undergoing vertical erosion include a large bed load comprising coarse hard particles. Potholes are common, as are deep narrow gorges.
- **Lateral erosion** increases a river's width. A large sediment load has to be entrained for this process to work most effectively; it is responsible in conjunction with the processes of slope transport and mass movement for valley widening, meander migration and river cliff formation.
- **Headward erosion** increases the length of a river. This process is most active in the source area of a river or where a bed is locally steep, this causes accelerated erosion and is commonly associated with waterfall formation.

Actual erosional work is then carried out by a number of processes.

- **Corrasion and attrition** These two processes rely upon the load of the river to achieve their effects. Corrasion occurs most often during periods of higher river flow, bed load being used as an abrasive agent, scratching and scraping at the solid bedrock. A correlation exists between this process and accelerated vertical erosion. The debris that results from the corrasive processes is free to collide and bash into itself. This process is known as attrition (or communition) and causes a reduction in particle size in a downstream direction. Smaller particles ensure that the channel remains smooth and that friction does not compromise efficiency in the lower course.

- **Hydraulic action** Water alone is not as effective an agent of erosion as a river 'charged' with debris. Hydraulic action is least effective in areas of hard base rock. In the middle and lower courses, where the bed and banks are likely to be composed of incohesive sediments and where there is a degree of sinuosity, bed and bank scour is effective in removing material vertically and laterally. An extreme form of hydraulic action is **cavitation**; the sudden and violent collapse of bubbles created by this process shatters banks extremely rapidly.

Transport

For sediment to be moved the following must occur.

- Resisting forces have to be overcome.
- When drag and embedded particle *inertia* is overcome and the particle begins to move, this is called the **critical tractive force**.
- **Competent velocity** has to be achieved, this is the lowest velocity at which particles of a particular size are set in motion, i.e. the bigger the particle the greater the velocity needed to move it.

Methods of sediment transport in rivers

- **Traction** – boulders and rocks are rolled along the bed.
- **Suspension** – particles are held in the body of the water.
- **Solution** – material is dissolved in the water.
- **Saltation** – particles are bounced along the bottom.

A stream 2 m wide by 10 cm deep can move 1½ tonnes of alluvium to the sea daily!

Downstream changes in sediment

- Amounts of material moved increase in a downstream direction, as weathered material is input and as tributaries add material.
- Material gets progressively smaller and rounder on its journey downstream.

Deposition

Depositional processes occur when the carrying capacity of a river is reduced. Several factors can affect the capacity of a river to retain its transported load.

- Water velocity changes due to changes in gradient, a break in slope. This is usually caused by variations in geology.
- Geology can, by changing the chemical composition of river water, cause rapid vegetational growth, this slows water and causes deposition.
- Evaporation (as on the Nile) or over-abstraction (as on the Colorado) can reduce flow and consequently bring on deposition.
- Additional debris and water volume from tributaries can result in the slowing down of water and to deposition.

Thus deposition is not confined to the lower reaches of rivers: it can occur at almost any point along the river's course. This localised deposition can, during periods of high flow, cause localised flooding.

Fluvial features

AQA A	U1	EDEXCEL A	U1
AQA B	UA	EDEXCEL B	U1
OCR B	U1	WJEC	U1
		NICCEA	U1

This section examines the effects of river processes on the landscape.

The graded profile or profile of equilibrium

Different parts or sub-sections of a river's course have different characteristics.

The graded profile

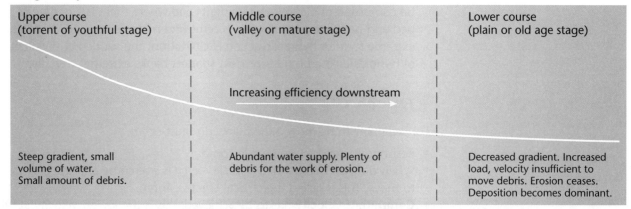

Upper course (torrent of youthful stage)	Middle course (valley or mature stage)	Lower course (plain or old age stage)
	Increasing efficiency downstream	
Steep gradient, small volume of water. Small amount of debris.	Abundant water supply. Plenty of debris for the work of erosion.	Decreased gradient. Increased load, velocity insufficient to move debris. Erosion ceases. Deposition becomes dominant.

The increasing efficiency is achieved through changes in the variables that influence the channel shape and form as shown below right.

The graded profile or state of equilibrium is thus achieved when the river's course is as efficient as it can be from source to mouth. Do remember though, a river course is rarely graded! Changes in rock type and sea level ensure the profile is constantly changing!

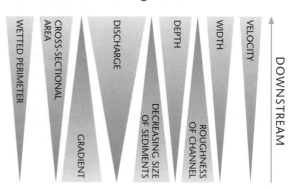

> You know the processes – now learn the 'features' – especially for the structured papers!

The variation in gradient, volume of water and amount of debris in the different stages of the river from source to mouth lead to the development of characteristic landforms in the different river stages.

The upper course

Features include:

> It is important you can draw diagrams to support your knowledge of landforms, waterfalls, gorges, etc.

*   **Interlocking spurs** See diagram, e.g. River Dane, Derbyshire.
*   **Potholes** Where pebbles and debris swirl around in joints and hollows on the river bed, gradually drilling a hole in the river bed, e.g. River Taff, Glamorgan.

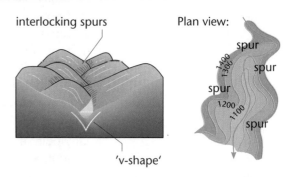

*   **Waterfalls and rapids** Common in the upper course, where there is a variety of different strength rocks, steeper gradients and fast water.
*   **Gorges** Over the course of time waterfalls and rapids migrate upstream. This forms a dramatic transverse profile, called a gorge.

The middle course

The change in the shape and size of the valley is due to the way water flows through meanders: erosion occurs on concave banks, deposition on convex banks. Meanders move downstream.

Cross-section through a meander

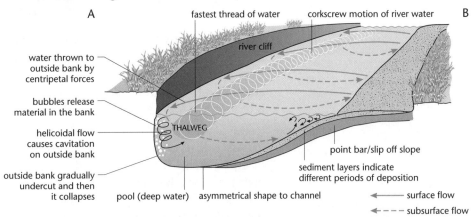

- **Meanders** The principle feature of the middle course. It is not clear how they form, but there are a number of misconceptions as to their initiation. Obstructions are unlikely to initiate meanders, it is more likely that the deformation of the river bed holds the key. Meanders may or may not be attempting to release excess energy from the river system; e.g. River Ribble, Lancashire; River Yare, Norfolk.

Plan of a meander belt

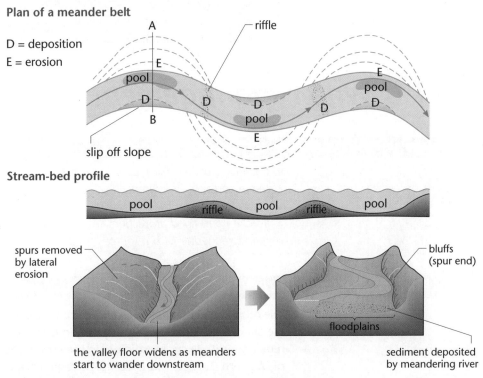

The meanders continue to grow in size, the floodplain gets bigger. Bluffs are cut. When the river floods it leaves material (deposits) over a floodplain.

The lower course

Features include:
- **Ox-bow lakes** As meanders grow in size they can breach their banks. When this happens a cut-off/mort lake or ox-bow forms, e.g. Mort Lake, London and the Trent valley, Nottinghamshire.

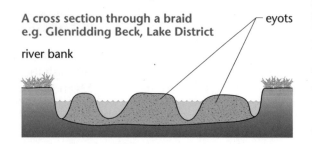

Braid formation

channel side — eyots

fluvial deposition

anastomosing channel

A cross section through a braid e.g. Glenridding Beck, Lake District

eyots

river bank

- **Braiding** occurs when a river flows or moves through a series of interlocking channels, rather than through a single thread. They are thought to form because of load *vs* discharge differences, induced by changes in slope or additions of water from tributaries/floods. They are highly unstable as the river is trying to achieve a more efficient profile (see diagram above).

- **Deltas** – the biggest of the world's rivers reach the sea, a lake or a lagoon with a massive load of material. This debris is dropped into the calmer water of the receiving areas. The salty water flocculating debris also aids the formation of deltas, as does the shallow angle of the coastal strip. Three main types of delta exist: **arcuate** e.g. the Mekong Delta, SE Asia; **birds foot** e.g. the Mississippi delta, USA; **estuarine** e.g. the Seine delta, France.

> Other features include: river terraces, meanderbelt.

Wide floodplain. e.g. River Cuckmere, East Sussex

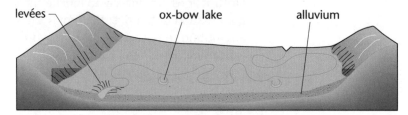

levées   ox-bow lake   alluvium

1.3 Water management

After studying this section you should understand:

- *the effects of variable regimes in rivers*
- *the importance of hydrographs*
- *human effects on the river basin*
- *flooding and flood protection*
- *managing water as a resource*

LEARNING SUMMARY

Variable regimes in rivers

AQA A	U1	EDEXCEL A	U1
AQA B	UA	EDEXCEL B	U1
OCR A	U1	WJEC	U1
OCR B	U1	NICCEA	U1

The regime of a river refers to variations that occur seasonally in discharge. Regimes are affected by climate. For instance, in tropical/equatorial climates there may be a regular or simple cycle exhibited in the regime. In the 'seasonal' temperate climates of Western Europe more complex regimes are common with 'multiple' peaks, affected variably by snow/glacial melt, rainfall and evapotranspiration.

> Know in detail the importance of hydrographs for your AS exams.

Hydrographs

For single storm/precipitation events the relationship between precipitation and discharge are shown on a **storm hydrograph**.

The appearance and nomenclature (labelling) of the hydrograph are shown opposite. Various factors can affect its shape. Physical variants, such as the nature of the inputs (principally precipitation) and the characteristics of the catchment and shape, rock type, relief and deforestation/afforestation can cause changes in the storm hydrograph shape. (See hydrograph diagrams below.)

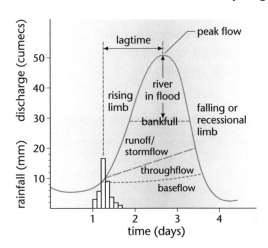

Some definitions and influencing factors.

- Hydrograph size and shape;
    high rainfall   = greater discharge.
    big basin   = greater discharge.
    elongated basin   = steady discharge.
    large symmetrical basin = flashy discharge.
- Lag-time; is the time interval between the peak of the rainfall event and the maximum discharge. It can be affected by channel steepness and drainage basin shape.
- Peak-flow; is greatest in the very largest drainage basins. Steep mountainous catchments cause high peaks. Lowland catchments have flatter peaks.
- Baseflow; maintains river flow away from flood periods.

How three factors affect hydrograph shape

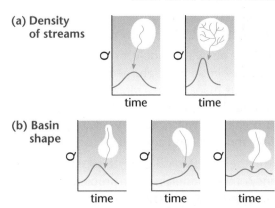

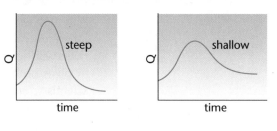

Human activities

These have direct and indirect effects upon the hydrology of a basin: the input of water to a river channel by land drainage, the storing of water in reservoirs increasing evaporation and regulating discharge; changing patterns of throughput by deforestation decreases interception and evapotranspiration, and increases overland flow. Urbanisation which creates impervious tarmac surfaces preventing infiltration, increasing overland flow and resulting in shorter lag times and higher discharges. Humans affect outputs by irrigation and water abstraction, and induce long-term changes in groundwater supplies.

Show your knowledge of topical geography here. Use up-to-date examples and collect newspaper articles of actual events.

Floods

Floods are the most important aspect of river flow. When rivers are in flood they have the greatest energy, changing the landscape dramatically. Floods affect man. They destroy property and crops, and kill! In the UK floods are relatively uncommon, recurring as they do every two to four years. They generally result from a coincidental combination of a number of factors or conditions (i.e. dry/bare ground and torrential rain). The more serious flood events are attributable to the fact that man has always been attracted to low-lying land; nowadays we are prepared to take the risk that protective measures will save our property, and us!

Small basins are especially prone to flash flooding. 80% of lives lost to drowning occur in such floods.

Know some case studies, e.g.:
- Rottingdean
- Badajoz, Spain
- 'Great Flood' of San Antonio, Texas, 1998.

Flood prediction

Man's predisposition for settlement sites on floodplains with the consequent loss of life, communications, farmland and industry make flood prediction a necessity. In the UK hydrologists working for the Environment Agency are responsible for flood prediction.

They use three main approaches:

- flood magnitude analysis based on measured characteristics, on flood hydrographs
- recurrence interval calculations based on computer stored and calculated statistics (recurrence is to do with the interval between similar sized floods)
- probability predictions (useful but not reliable).

Our response to floods

Know examples of flood control schemes in the UK. Many are well-documented.

Response	Control	
Adjustments, *in actions on the floodplain*	Abatement, *of problems in the catchment*	Protection, *along the channel*
None	Afforestation	Walls and embankments
Emergency action	Change agricultural practices	Channel improvements
Flood proofing	Change the vegetation use	Diversion schemes
Land use regulation	Affect of urban areas	Reservoirs
Financial disincentives		Barrages; Flood barriers

Water as a resource

AQA A	U1	EDEXCEL A	U1
AQA B	UA	EDEXCEL B	U1
OCR B	U1	WJEC	U1
		NICCEA	U1

There is not an infinite amount of water in the world – we have to manage it. Many AS questions focus on this topic.

Although we can no longer take our water supply for granted, most river systems are inherently robust (for instance, salmon have returned to the Thames). Nevertheless, rivers will increasingly need to be managed in a more sustainable way. Management involves both the present and the future. Geographers, biologists, environmental scientists, engineers and politicians will need to work together to ensure water supply for our consumption, industrial and agricultural use.

Man will have to interfere in the hydrological cycle, to supply water for our growing populations. Our occupancy of the floodplain is also problematic. We have to control the excesses that occur in rivers and maintain and regulate flow to ensure water quality, to allow for transport and leisure, and to maintain and enhance the landscape.

Management issues

Environmental solutions to flooding

Since the end of the nineteenth century we have 'trained' rivers, mostly to control flooding. Such training is expensive, extensive and unnecessarily manipulative. This area of management has seen the most promising of reforms over the last decade. Many environmental solutions are now used to control floods and flooding events.

The effect of hard engineering, from the past, can be controlled 'naturally', by returning the river to its former state.

- Pools, riffles and meanders are now reconstructed, as they are seen as the best, most stable alternative to artificial straightening.
- The re-introduction of vegetation is seen as important, as it duplicates and promotes the bank stability of the natural stream channel.
- Stream maintenance, which is small scale and in harmony with the watery environment, is seen as important.
- Bio-technical methods copy and reproduce the natural symmetry/asymmetry of the stream channel.

Pollution control

Some of the most polluted river courses are those near roads and housing, with heavy metals and other sediments washed into the water.

The flush toilet and liquid waste from industry have both contributed to the deterioration of our water systems, and nitrates and herbicides (in the UK) have accelerated the problem in the last 25 years. Legislative development, spurred on by European initiatives, has built apace in the last twenty years. Rivers are now cleaner than they have been for hundreds of years. We are much more aware of the potential to pollute groundwater, of the relationship between rivers and marine pollution and for the need for more integrated pollution control. Pollution control has clear long-term benefits and will continue to be scientific and policy biased in the future.

The main sources of groundwater pollution are landfills, chemical processing industries, gas works, power and petrol stations.

A future for groundwater?

200 years after the industrial revolution, we still rely on groundwater as an economic source for our urbanised population. However, in some areas groundwater in the main aquifers is being depleted and polluted. Greater attention will have to be devoted to the management of aquifers in the future to ensure effective supply and use for drinking water, and as part of an integrated system using rivers, reservoirs and groundwater. Contamination from agricultural land, from landfill and from industrial seepage must be controlled, as must the maintenance of adequate river flows.

Disputes and dam building

Ideal conditions for dam construction:

- Upland area
- Away from population
- High precipitation area
- Impermeable rock
- Narrow, steep-sided valley for dam construction.

There are also environmental, economic and political considerations.

The disputes that surround dam construction are complex and these involve environmental, economic and political factors. Most dams that are built tend to be multi-purpose, supplying energy, providing water for irrigation and allowing for recreation and leisure pursuits. Hydro-politics, or the management of water resources in shared basins, is the most contentious of issues related to dam building at the present time. Interference with water in an international river always has a downstream affect that can lead to intense confrontation.

Damming of the Euphrates and Tigris Due to be completed early this century, the ambitious South-eastern Anatolian Project will capture a significant proportion of the headwaters of the Tigris and Euphrates. The biggest dam, the Ataturk Dam, is at the centre of the project that will create 22 dams. The purpose of the scheme is to improve the economic base of SE Turkey and to provide a springboard for industrial and agricultural development. An obvious consequence of this damming is that the downstream lands of Syria and Iraq are being deprived water and Kurdish lands are being flooded. Worth noting is that Turkey has no legal obligation to provide water downstream of its borders! This water piracy is expected to heighten tension in the Middle East as the Euphrates flow is reduced. Hydro-politics have also influenced the building of the Three Gorges Dam (China), the National Water Carrier Scheme (Israel), the diverting of the Danube (Czechoslovakia and Hungary). Hydro-politics will continue to influence dam building through the next millennium.

CASE STUDY

Water transfers

A 'national water grid' has been suggested for the UK, to help during droughts.

Under the 1991 Water Resources Act, the Environment Agency must ensure that water resources are conserved, redistributed, augmented and properly used. Demand management is at the forefront of the strategy, ensuring the optimum use of the existing resources. One way that resource delivery has been improved is through schemes to transfer water, a costly option but one that will increasingly develop after some two decades of overbuilding of dams.

Examples of transfers in the UK
- Kielder Water, UK, was built to cope with expected demand for water by industry on Tyneside. By the mid-eighties the expected development of industry had still to take-off. One useful development related to the Kielder project was the building of a water transfer viaduct to the Tees, to the south of Kielder and the River Tyne.
- East Anglia, UK. In East Anglia the Ely/Ouse/Essex transfer scheme was completed by Anglian Water in the 70s to move water from the north of the region to the burgeoning populations and industries in Essex.
- Lake Vyrnwy, in Wales has supplied Liverpool, 109 km away, for the last 100 years.

CASE STUDY

Sample question and model answer

(a)

The diagram below shows a stream channel cross section. Values for the stream velocity (in metres per second) are given for the locations indicated by the dot for the decimal point.

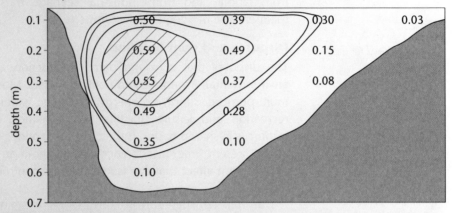

Most structured questions require the completion of graphs/diagrams. Such skills need practice.

This question enables you to reveal the thalweg (the fastest thread of water) in the stream using a technique called interpolation. This is like drawing contours; a 'contour' interval is chosen and lines are carefully drawn on, in pencil.

(i) On the diagram shade in the channel area in which the measurements suggest there is a stream velocity over 0.50 metres per second. [3]

The shaded portion reveals the position of the fastest thread of water. Accurate interpolation has located this fast water.

(ii) Explain how the differences in velocity shown affect the processes operating in the stream channel at this point. [4]

The answer is looking for four good points!

- *Velocity is greatest on the outside bend. Lateral erosion, hydraulic action and abrasion occur.*
- *Velocity is lowest on the inside bend. Deposition occurs, competence and carrying capacity are reduced. Frictional effects are obvious.*
- *Helicoidal flow causes cavitation on outside bank.*
- *Greatest transport of sediment is in suspension, linked to the greatest velocity thread.*

(b)

For **one** named river basin, locate and explain the formation of those channel features which have resulted from vertical and headward erosion by the river. [8]

Here is your chance to demonstrate the effects that velocity has on stream channels. Candidates will link their theoretical process knowledge to this new situation.

This question obviously relies on recall – but to get an 'A' grade it is equally important to produce a logical development of ideas and discussion.

As part of our studies we have looked at the River Dane in Derbyshire, it shows all the features of an upland river. The valley is steep sided, even gorge-like in places, with a boulder strewn bed. Outcrops of millstone grit and limestone are frequent. Potholes are common. These features are typical of a river that has undergone and is undergoing vertical erosion i.e. it is cutting into its bed. Headward erosion, has caused the river to retreat further upstream, a link perhaps to changes in base level. In places small rapid and waterfall formations have appeared, these make so-called knickpoints. My studies in and around Gradbach have helped me to understand more fully the processes of headward and vertical erosion.

Allocate your time according to marks available – but allow your ideas to 'flow'. Plan your answer on the margin of your paper!

Examiner's note: the question looks for discussion of:

- *possible features e.g. waterfalls, rapids, gorges and potholes,*
- *locations e.g. actual and upper course,*
- *explanations of vertical and headward erosion.*

AQA (modified)

Practice examination questions

Section A

1 Study the diagram below which shows a storm hydrograph.

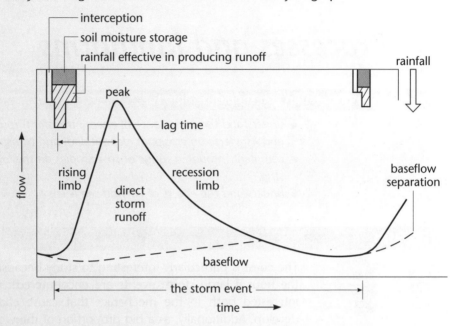

(a) For the period of the storm event, explain the form of the graph. [4]

(b) State **three** factors which affect the rate of interception. [3]

(c) Why is baseflow relatively stable? [2]

(d) (i) State **two** catchment features which affect lag-time. [2]

   (ii) for both features, describe how the lag-time is affected. [4]

(e) How might a single rain storm produce a hydrograph with more than one peak? [2]

(f) Comment on **two** ways in which rain storms might affect channel form. [4]

(g) What information from hydrographs might assist flood control in a river basin? [4]

Edexcel

2 Section B

Explain why both human and physical factors are studied to explain size and frequency of river floods.

Illustrate your answer with reference to one or more examples. [15]

AQA

Chapter 2

The challenge of the coast

The following topics are covered in this chapter:

- *Processes and landforms*
- *Management and planning*

2.1 Processes and landforms

After studying this section you should be able to:

- understand the importance of energy transfer through the coastal system and its effects on erosional, depositional and transportational processes
- comment through a range of case studies on the landforms of the coastal strip
- understand the effects of varying sea levels

LEARNING SUMMARY

Processes

AQA A	A2	EDEXCEL A	U1
AQA B	UB	EDEXCEL B	U1
OCR A	A2	WJEC	A2
OCR B	U1	NICCEA	A2

The coast is particularly interesting to study because it is constantly changing (by the hour if large storm events are encountered). Coastal geomorphologists are interested both in the mechanics that cause change and the landforms that develop. Additionally, as a big proportion of the world's population lives near the coastline we have to deal with the threats and problems posed to human habitation. Problems include: flooding, rising sea levels, accelerated erosion, the effects of industrial pollution and the effects of tourism. Careful, sustainable management of the coast helps to deal with these problems.

The coast is defined in geographical terms below.

> You need to know these terms.

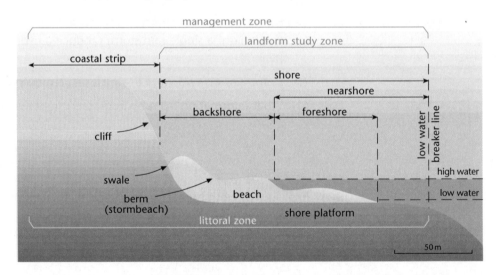

> They can also be sub-divided based on whether they are an emergent/submergent primary/secondary or high or low energy coast.

Traditionally, the approach adopted by coastal geomorphologists has been to try and classify the coast, according to whether the dominant process is erosional or depositional. However, little to date has been achieved. Currently the relationships of mechanisms and processes to landforms are again being explored, the so-called 'process–landform response approach', below.

> Note: 'sediment transport' is a coverall for erosion + deposition. There is no permanent loss of sediment, it is just moved back and forth.

Inputs of energy	→	Processes	→	Landform response	→	Outputs
Tides, wind and waves		Sediment transport		2d – beach (slope/shape) 3d – landforms (stacks/cliffs etc.)		of energy (breaking waves) and sediment on the sea floor

Waves, currents and tides

The interaction between waves, tides and coastal currents shape, modify and mould the shoreline. Waves provide the energy for the coastal system, tides spread energy over a larger vertical area of the coast, and currents spread and redistribute energy/sediment along the coastline. All three can act against and work with one another.

Waves

Remember: wave size is determined by the fetch (the distance over open water that the wind has blown) and the strength and duration of the blow. N Norfolk has a fetch of over 1500 km.

The force behind the formation and shaping of the coast is the wave. The drag effects of wind across the sea cause undulations on the surface. As these undulations build (because of pressure contrasts on the windward and leeward sides) so water starts to move in an orbital or oscillatory fashion inside them. This movement is related to their height. Energy conversion, from potential to kinetic, occurs continually within these waves. Waves are, then, a means of moving energy through water with only small displacements of water particles in the direction of energy flow.

Waves that break can be either **constructional** or **destructive**.

(A)   **(B)**

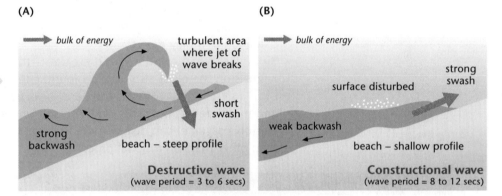

Swash runs up the beach. Backwash runs down the beach.

Four types of wave have been identified; surging, collapsing, plunging (as in A) and spilling (as in B).

As waves move nearer the coast, the submarine contours/sea bed starts to affect the waves. Locally both friction and drag start to increase. This is known as **wave refraction** and has the effect of varying the available energy along a coastline.

Currents

Shore-normal currents establish a cell circulation in the near-shore zone. A large amount of sediment is moved up the beach by the swash and is balanced by the rip of the backwash running back down the beach. The vast amount of water moving back down the beach forms a **riphead**, a deep (up to 3 m+) energy/water dissipation hollow offshore. The effect of the strong, sediment-laden swash is to form a beach **cusp**, the smallest of the beach depositional features. (It is composed of graded/sorted sediments in a horned cusp.)

Waves approaching at 30° move most sediment along a shore!

Most sediment on beaches has a river or offshore origin.

Most sediment moves W to E in south of the UK. Along East Anglia's coasts it moves N to S.

For the most part waves approach coasts at a slight angle, less than 10° is normal. These wave-normal/oblique currents aided by winds and submarine currents carry sediment up the beach at the same angle as the wave/current, and return it in 'rips' perpendicular (at right angles) to the beach. The net effect is to move material along a beach; this longshore movement is called **longshore drift** (see diagram).

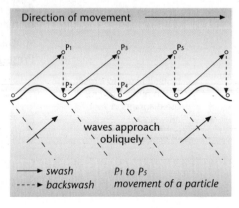

Tides

Erosional processes are concentrated between high (HTM) and low tide (LTM) marks, in the splash zone above the HTM and the wave base below LTM.

Erosion and subaerial weathering

Three main erosional processes work to destroy the coastline:

30 tonnes pressure/m² of 'explosive' effect.

- **Quarrying** (on hard rock coastlines – granite), or **hydraulic action** (on soft rock coasts – sandstone, glacial till). The compression, as a wave hits a coast, and expansion as it retreats exerts considerable pressure on the coastline.
- **Abrasion/corrasion** – waves use the pebbles and cobbles to corrade the cliff base, undercutting it rapidly. Weaknesses in rocks, joints, etc., are differentially exploited by this form of erosion.

Can be heard working on pebbly beaches!!

- **Attrition** – reduces the size, and rounds off individual clasts (pebbles) as they bash into one another.

Subaerial processes include:

- **Corrosion and solution** – the chemical dissolving of rock in acido-saline conditions.
- **Sea weathering** – slaking caused by the alternate wetting and drying of the coast, along with salt crystallization are most prevalent. But hydration and oxidation also play a part.
- **Biological weathering and bio-erosion** – the effects of boring plants and animals on the coast also have profound effect.
- **Mass-movement** – rockfalls (on hard rock coasts e.g. North West Scotland) and mudslides, slumps and slides (on soft unconsolidated rock, the tills of the Norfolk coastline) can deliver more material to the coastal sediment system than other marine erosional processes.

Landforms

AQA A	A2	EDEXCEL A	U1
AQA B	UB	EDEXCEL B	U1
OCR A	A2	WJEC	A2
OCR B	U1	NICCEA	A2

Note: different combinations of height, orientation, steepness, determine cliff height.

A relationship between rock type, erosion and cliff morphology exists.

Cliffs

Cliffs are probably the principal and most obvious feature on the coastline: they vary in height, orientation, steepness; they vary in terms of their lithology and structure. For instance, the cliffs on the soft boulder clay/till of North Norfolk rarely exceed 8 metres or so, whereas the Cumbrian cliffs of St Bees Head reach 25 metres.

Wave energy concentrates energy at the cliff base, forming a wave-cut notch, the size of which is determined by the tidal range. As time passes the overhanging cliff eventually collapses, this material is used as ammunition to accelerate the creation of the wave-cut notch.

Shore platforms

Note: if asked to explain beach profiles you must use diagrams in your answer.

Know your processes!

As shorelines, coastal slopes and cliff lines are eroded there is a marked/obvious retreat or recession that leaves behind a platform. These platforms have an overall convex shape and an average slope of 0° to 3°. Most show a break in slope, marking the LTM. Present-day processes have produced most platforms. e.g. off Bembridge, IOW

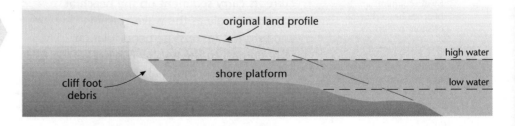

For a long period of time a single process, abrasion, was held responsible for the formation of shore platforms. The modern view is that shore platforms have a multiple-process origin. They may be caused by a combination of the processes below.

- **Abrasion** – sand grains moved by waves plane the platform surface. This is particularly effective in the upper shore section.
- **Mechanical wave-erosion** – the process of quarrying, through wave hammer, compression or pressure release, picks out and exploits variations in lithology. This process causes cliff recession and roughens the platform surface.
- **Weathering** – wetting and drying can cause hydration and oxidation, and salt crystallisation. These processes are slower than the violent processes outlined above.
- **Subaerial processes: solution** – the chemical solution of calcareous rocks (limestone/chalk) in tropical areas is a recorded phenomenon. Even Norfolk's chalk shore platforms are affected by solution; slight rises or falls away from the atypical sea temperature produce rapid chemical stripping of the submerged platform.
- **Tides** – vary the level of process activity.

Other features of coastal erosion

A note on concordance and discordance.

Concordant or longitudinal – where the 'grain' of the country rock is parallel to the coast. e.g. Lulworth Cove.

Discordant or transverse – where the 'grain' runs at right angles to the coast. e.g. Swanage.

The grain reflects the structure – the alignment of the folds or the different types of rocks. This leads to **differential erosion**.
A Dorset, Lulworth Cove and Swanage case study follows.

Arches and stacks

- The sea erodes along a line of weakness (e.g. a fault) in a headland to form a cave.
- Caves formed on opposite sides of headlands join to form an arch.
- Arch will eventually collapse to form a stack.

Geos and blow-holes

Where a fault in a cliff at right angles to a coast is eroded by the sea, a long, narrow inlet may form, e.g. Huntsman's Leap, Pembroke. The first stage may well be a cave which connects to the surface by a chimney to form a blow-hole or gloup.

Some physical features and processes found along the Dorset coast

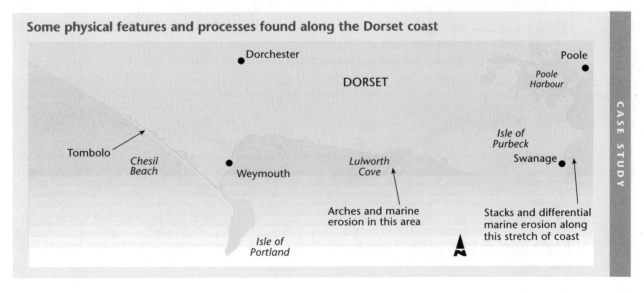

Pronounced headlands alternate with wide bays where the softer clays have been more easily eroded. Wave refraction concentrates erosion on the headlands. Extensive sandy beaches have accumulated in the shallow and sheltered waters at the head of the bays.

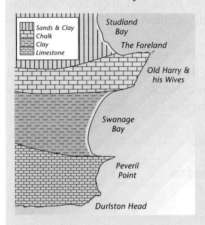

DORSET

West Bay
ridge 7m high
small particles

drift of large and small particles with strong longshore current

The Fleet

Chesil Beach

Weymouth

LOW WAVE ENERGY

Lyme Bay

drift of small particles only with weak current

ridge 13m high large particles

HIGH WAVE ENERGY

Isle of Portland

10 km

Sands & Clay
Chalk
Clay
Limestone

Studland Bay

The Foreland

Old Harry & his Wives

Swanage Bay

Peveril Point

Durlston Head

Limestone has almost completely gone, a few isolated rocks remain at low tide. Durdle Door is a fine natural arch.

At Stair Hole the sea has broken through the Limestone (via caves and joints) and is beginning to attack the soft Wealden clays. Lulworth Cove is a near circular bay, extending E to W, its growth impeded by chalk. Its growth was initiated by the stream flowing into the sea through the limestone.

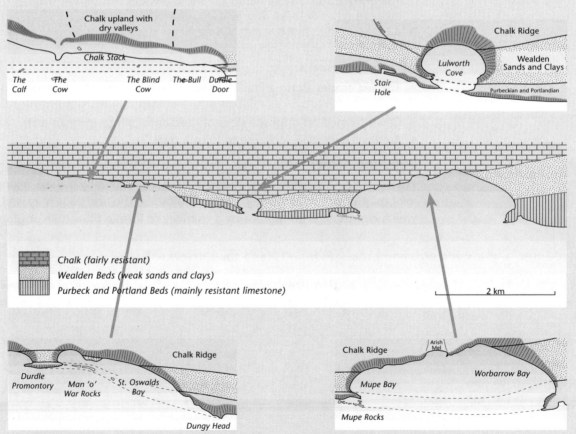

Chalk upland with dry valleys

Chalk Stack

The Calf | The Cow | The Blind Cow | The Bull | Durdle Door

stream

Chalk Ridge

Lulworth Cove

Wealden Sands and Clays

Stair Hole

Purbeckian and Portlandian

Chalk (fairly resistant)
Wealden Beds (weak sands and clays)
Purbeck and Portland Beds (mainly resistant limestone)

2 km

Chalk Ridge

Durdle Promontory

Man 'o' War Rocks

St. Oswalds Bay

Dungy Head

Chalk Ridge

Arish Mel

Worbarrow Bay

Mupe Bay

Mupe Rocks

St. Oswalds Bay is an elongated double bay, formed by the amalgamation of two former coves. The Man 'o' War Rocks are all that remains of the limestone.

The small but pronounced bay at the dry gap at Arish Mell is in the early stages of breaching the next resistant barrier. The Mupe Rocks are stacks.

Landforms of coastal deposition

The debris from coastal erosion is moved and deposited by waves and currents. Major depositional landforms are shoreline beaches and detached beaches (spits and bars). Each landform has its own dynamic sediment store/budget, with material also being lost and gained.

Beaches

These are perhaps the most widespread of depositional landforms. They are geomorphologically successful because of the mobility of their loose sand sediment. These complex systems exhibit a range of minor landforms, shown below.

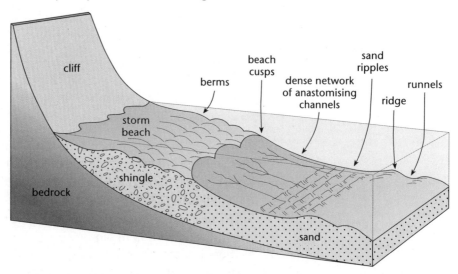

Most of the landforms shown above are the results of coast processes reworking beach sediments. As the best form of defence for a coastline (as they absorb energy) they are easily destroyed by natural and man-induced processes (see later management section).

Detached beaches

The two major detached features are the bar and spit. The following account incorporates a variety of examples to explain the formation of these detached features.

- **Bars** Created by the action of breaking offshore waves on gently sloping shores. The breaking waves excavate material from the sea floor to form submarine bars, which slowly build up until they appear above SL; many are frequently quickly vegetated. Bars frequently move ashore/inland, trapped sea water in the form of a lagoon is filled by sediment and marsh vegetation; sand dunes tend to be the final relic of the bar (e.g. Looe, in Cornwall). The best examples of bars in the UK are found off the Norfolk Coast between Hunstanton and Sheringham, where the sea floor consists of unconsolidated glacial material. The shingle formations exhibit a wide variety of features due both to straightforward wave action and longshore drift. The usual form in Norfolk is crescentic; drifting that takes place along the seaward side leads to modification and complication. Scolt Head Island is a fully developed offshore bar, off the North Norfolk coast.
- **Spits** Created for the most part by longshore drift, they are attached to the shore and end in deep water. The diagram shows how Blakeney Point Spit may have formed the typical features of a compound spit and it is likely that Blakeney started its life as an offshore bar!

Other depositional features to be aware of are tombolos (the mode of formation of Chesil Beach) and cuspate forelands e.g. Dungeness Foreland.

Features of Blakeney Spit

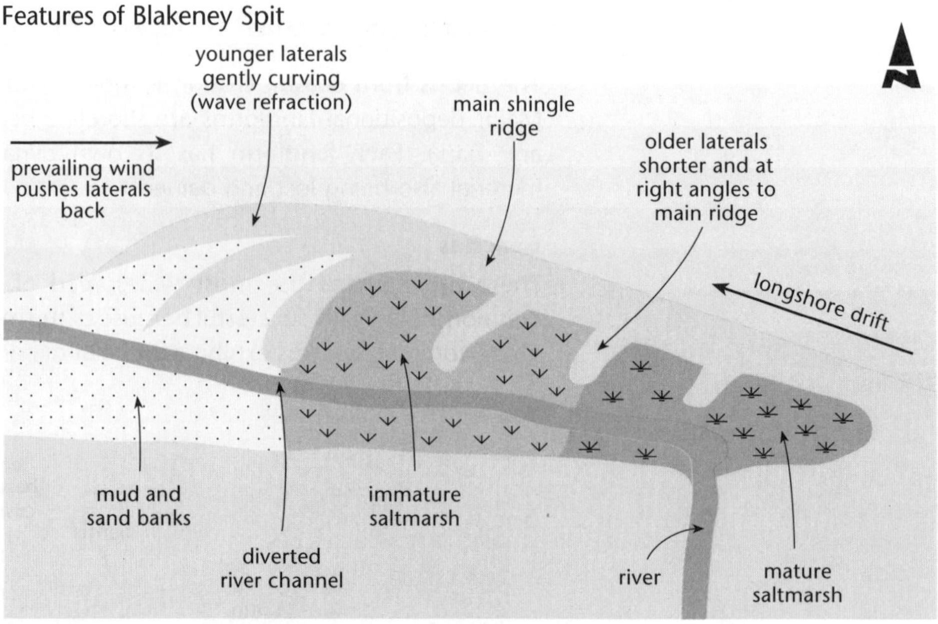

- **Tombolos** Chesil Beach (see page 38) is a fine example of a tombolo, a 30 km shingle ridge connecting the Isle of Portland to the mainland. From West Bay to Abbotsbury it hugs the coast; then the elongated lagoon – the Fleet – separates it from the mainland. It has two unusual features:
  - The beach is a simple ridge reaching a maximum size at Portland, where it is 60 m wide and 13 m high.
  - There is a progressive grading in the size of the material. At West Bay the shingle is small, about the size of a pea; by Portland it reaches the size of one's fist.

  Whether Chesil beach is a spit or a bar driven ashore is not clear.

2.2 Management and planning

After studying this section you should understand:

- *that the coastal strip is used extensively by man*
- *that the coast has to be protected and preserved*

LEARNING SUMMARY

Managing coasts

AQA A	A2	EDEXCEL A	U1
AQA B	UB	EDEXCEL B	U1
OCR A	A2	WJEC	A2
OCR B	U1	NICCEA	A2

Analysis of the last census shows us that 16.9 million of the UK's population lives within 10 km of the coast. Buildings, roads and recreation facilities occupy 31% of the coastal frontage of the UK. 40% of the UK's manufacturing industry is also situated on or near the coast. Economic pressure for further expansion of these facilities is likely in the future. With so much of the coast developed or being developed, there is a demand for coastal protection from erosion, and sea and tidal defences against flooding. Coastal defences have to protect the population and their economic well being. Approximately 2100 km of coast defences presently defend our coastlines. The bulk of the coastal defences are along the east and East Anglian coast. Management of our coastlines is the responsibility of Maritime Local District Councils, MAFF and the Environment Agency (and other groups). All defence schemes need planning permission and consultation before building can begin.

Our coast is vulnerable – know how it's protected.

Coastal protection

	Cliff face strategies	Cliff foot strategies	Beach management schemes
Hard engineering	Cliff pinning Cliff modification Drainage of cliffs Gabion baskets (Average costs £500 to £2000/m)	Sea walls (£1500 to £2000/m) Revetments (£300/m) Rip-Rap	Breakwaters (£120 000 each) Groynes (Wooden at £6000 each) Beach Pumping Reef Systems (£ millions)
Problems associated with hard engineering	Over drying and subsistence. Gabions are useless as an erosion preventer.	Expensive to build and maintain. Walls cause accelerated erosion of the beach and allows beach levels to fall. Revetments have a short life cycle.	Groynes interrupt and reduce LSD as they are 100% efficient at trapping sand. They are visually intrusive. Reefs change beach plans and profiles.
Soft engineering	Revegetation	No engineering problems	Beach nourishment and replenishment (at a cost of £20/m^3)
Problems associated with soft engineering	Problems only relate to poor vegetation choice.	No problems	Near-shore dredged material can affect the sediment cell, impeding and disrupting replenishment. Recharged sediment needs to be of the same calibre to the natural sediment.

KEY POINT

In the UK this generally involves hard (structures built to resist wave energy) or soft (solutions that work with the environment and natural processes) engineering.

Happisburgh, Norfolk

With an excess of 70 000 m^3 of material being lost from the Happisburgh area every year, up to 6000 h of residential, agricultural and commercial properties were being put at risk. It was decided that nine shore-parallel reefs would be built to reduce losses. Completed in 1997 each reef is 250 m long, 45 m wide and contains thousands of tonnes of imported rock. Since completion the scheme has been beset by problems, most controversial being the formation of embayments along the coast. That is, those areas between reefs experience no real protection at all!

CASE STUDY

Sediment budget disruption

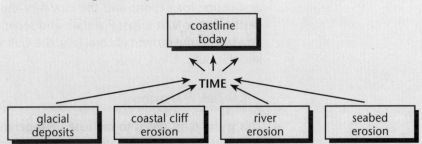

Disruption of the above sediment budget, usually by dredging, can have catastrophic effects for coastal communities...

A. Hallsands, Beesands and Torcross: offshore dredging for shingle for the building of Plymouth dockyards removed the only defence, an offshore bar, which these communities had. Much expensive, remedial work has been undertaken to protect these communities.

B. Off Southsea, Portsmouth: dredging of shingle for navigation purposes and as a metal for the M3/M27 extensions caused massive, though remedially repairable, damage to Southsea beach and promenade.

CASE STUDY

Planning and funding

AQA A	A2	EDEXCEL A	U1
AQA B	UB	EDEXCEL B	U1
OCR A	A2	WJEC	A2
OCR B	U1	NICCEA	A2

Once a coastal issue has come to the fore, a number of factors have to be taken into consideration before deciding the best management scheme for a coastline.

All Councils with responsibilities for coastal defences have strategies, **Shoreline Management Plans (SMPs)**, in place based on the so-called **Sediment Management Cells** (a way of dividing up the UK and Welsh coastline). These SMPs will identify the options available to the Coastal Managers and are known as **Management Units**, they are:

> Note the relationship between urban areas and thinly populated areas.

- do nothing – no action is taken to build or maintain defences
- hold the line – interventional to hold defences where they are at present
- advance the line/change – new measures that move defences seaward
- managed retreat.

After a problematic coastal issue has been identified and action appears sensible, funding is criterion tested. The first stage is to convince MAFF, and other agencies too, that the proposed scheme will be feasible in terms of the engineering, viable economically, and that it is environmentally sound. Once past the three E's test it is criterion scored, based on priority, urgency and economics. Only those schemes that score in excess of 22 out of 30 are even considered for funding. The result of this type of scoring is to favour urban areas over rural areas.

Coastal concerns

AQA A	A2	EDEXCEL A	U1
AQA B	UB	EDEXCEL B	U1
OCR A	A2	WJEC	A2
OCR B	U1	NICCEA	A2

Coral reefs and atolls

These are widespread between latitudes 30°N and 30°S in the western parts of the Pacific, Indian and Atlantic Oceans. They are either of fringing, barrier or atollic origin. They have a biological 'source', the remains of living and dead polyps, algae, foraminifera, molluscs and other shelly organisms in the presence of calcium carbonate contributing to the reef. Ideal sea conditions in which reefs form are where the salinity is in the order of 27 to 38 ppm, with a mean sea temperature of 18°C, and with an adequate circulation of sea water.

Changing sea temperature distributions (attributed to El Niño) and global warming (causing sea level changes) are affecting the continued development and growth of coral reefs. Destruction by human activity (tourism, over-fishing and wholesale destruction for export) and the crown-of-thorns sea star infestation has wrought destruction to vast areas. Pollution and sediment disruption also clouds water and slows/stops the growth of coral (e.g. the Gulf war and the release of oil by the Iraqis into the Gulf).

Global warming

This is the term given to increased temperatures on the Earth's surface, resulting from carbon dioxide and other gases trapping the incoming solar radiation. Global warming results in the world's oceans increasing their volume (so-called thermal expansion), a eustatic change. The predictions for SL as a result of global warming for the next fifty to a hundred years are alarming, the capital of Norfolk could easily be renamed Norwich-next-the-Sea! Seriously, 80 million extra people will be flooded each year due to rising sea levels.

Combined with an apparent increase in storminess, increased extreme tidal events and storm surges, lowland coastal areas will need enhanced protection from the sea.

How the sea might save us from global warming

It has been suggested that levels of carbon dioxide might be reduced by pumping iron sulphate into the sea. Plankton would thrive in this iron-rich environment and the increased numbers of plankton might absorb excess carbon dioxide.

Sea surges

These seem to be an increasingly frequent phenomenon. They result from a number of concurrent but freak conditions. North Sea surges, for example, result from:

- high tides
- strong northerly winds (influenced by the presence of low pressure)
- high pressure to the west of the North Sea and low pressure to the east
- the bottleneck that is the southern North Sea and its lowland coast exacerbating the problem (as does global warming).

Events in 1993, 1995 and 1999 caused flooding and damage to large parts of the Broads and coast of Norfolk, and parts of coastal western Europe.

Human impact on the oceans

Increased awareness of the environment has led to added pressure to safeguard our natural coastline and has contributed to land use conflict. The pollution of surface water and sub-surface water by chemicals released through industrial activities either directly into rivers or into the air, and by fertilisers used in agriculture, has led to pollution of coastal water, for example:

- Nitrogen phosphates – in sewage and fertilisers – lead to algal blooms, e.g. coastal Italy.
- Mercury, cadmium, hydrocarbons and DDT – from industrial activity and farming – lead to food chain disruption and human genetic problems, e.g. discharges from the River Rhine into the North Sea and Minimata Bay in Japan.
- Oil – from exploration and transportational accidents and from 'tank' cleaning – causes food chain disruption and faunal losses, and ecosystem crashes, e.g. Exxon Valdez, Sea Empress and Braer disasters.
- Chemical and radioactive – through waste disposal, e.g. into the Beaufort Channel in the Irish Sea.

Nobody denies damage is being done, disagreement centres on the severity of damage. Our coastline must be conserved, protected and enhanced.

Sample question and model answer

Sea level change is also due to changes in ocean capacity due to sediment infill.

Eustatic change is a worldwide change in sea level.

But what's the cause? It tends to be pressurised during glaciation and pressure released during de-glaciation. Geological down warping and tectonics also have an effect.

Good point.

The focus of the second part of the question is on humans not processes.

A plethora of information in 16 lines but it bears no relationship to the question set.

All the information above should have been in part (a).

Is this why they are used?

Is it more to do with available land as the soil tends to be skeletal and lacking nutrients.

They also provide natural routeways.

Accuracy?

Good: they are used as ports.

1

(a) Distinguish between isostatic and eustatic sea level change. [5]

Sea levels change in connection with the growth and decay of ice sheets. Eustatic change refers to a global change in sea level. The level of the land also varies in relation to the sea and these changes, which are more localised, are called isostatic sea level changes.   (awarded 2 marks on assessment)

(b) Examine the causes of and human responses to sea level change. [10]

AQA (AEB) (modified)

Eustatic changes in sea level occur mainly because of glaciation. As the glacial period began water was held on the land rather than being returned to the sea. The sea level fall is slow, with a rapid rise as the ice melts. Minor eustatic changes are caused by the ocean expanding in volume during warmer periods.

Isostatic sea level changes can be due to tectonic movement. In tectonically active areas the land rises faster than the eustatic rise of the sea e.g. in parts of the Mediterranean, and during earthquakes the shoreline can be elevated virtually instantaneously e.g. during the Chilean earthquake of 1835. Local sea level changes are also influenced by glaciation. Ice sheets press down on the land and so when they melt the land rises. Therefore landforms formed when the land was depressed become elevated forming raised beaches. For this to happen the isostatic uplift must exceed the eustatic rise of the sea. The amount of uplift depends on the thickness and weight of ice originally covering the area.

Raised beaches are used for agriculture because they are sandy and well-drained. Their position above the sea means that the conditions are less salty so some vegetation can grow. This allows soil to form and so they can be used for growing crops. Another use for them is golf courses, again because they are well-drained, but also because they often have good views and are attractive places.

A rise in sea level results in submergence of features. E.g. Fjords form where a glacier has eroded below the original sea level so that when the ice melts the valley is flooded by sea water. These can be found in Norway and their main function today is as tourist attractions. Rias are flooded river valleys which were eroded during the glacial period and are now used as deep water ports e.g. Salcombe Estuary, Devon.

Marks awarded; 2 for part (a), 4 for part (b).

Significant components were missed in both a and b sections and it was poorly planned.

AQA (modified)

Practice examination questions

1 Section A

Study the diagram below showing a plan of a coastal cliff and beach where marine transfer of sediment is active.

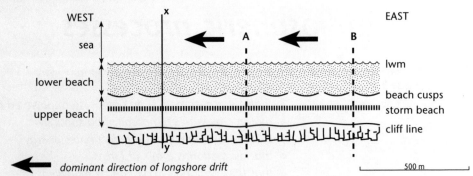

(a) (i) Sketch a labelled cross section to show the main changes in slope along the line XY on the above diagram. [3]

(ii) Name two factors which influence the steepness of the upper beach. [2]

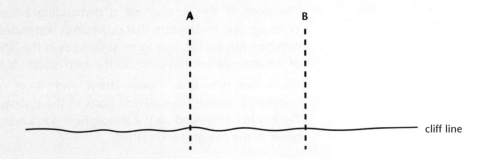

(b) (i) Draw on the diagrams above the likely plan of the beach about fifty years after the building of solid piers along the lines A and B. [2]

(ii) Explain the new distribution of beach sediment. [2]

(c) Discuss, with examples, the problems which can arise as a result of human intervention designed to influence coastal processes. [6]

Edexcel (modified)

Section B

(a) Distinguish between isostatic and eustatic sea level change. [5]

(b) Examine the causes of and human responses to sea level change. [10]

AQA (AEB 1997)

> Bear in mind the comments made about the student's answer to this sample question on p.44

Chapter 3

The challenge of the atmosphere

The following topics are covered in this chapter:

- Atmospheric processes
- Weather systems

3.1 Atmospheric processes

After studying this section you should understand:

- that the atmosphere is an open system powered by and circulating energy
- that energy varies temporally and spatially
- that the movement of air redistributes energy
- that air motion is a mix of forces
- that the circulation of air can be modelled
- that the small amounts of moisture in the air vary in time and space
- the basic mechanics of condensation, lapse rates and forms of condensation

LEARNING SUMMARY

Systems and energy exchange

OCR A	U1	WJEC	A2
OCR B	U1	NICCEA	U1
EDEXCEL A	A2		

The study of the atmosphere is demanding because of the complexity of the processes and mechanism that power, run and maintain it. For us as humans all the activities vital for life, and to most features in the landscape, occur in the thin layer of the atmosphere that clings to the Earth's surface because of the 'pull' of gravity.

The atmosphere is an open system, with flows or movements of energy and materials between the different parts of the system. The sun is the driving force. The key to understanding the atmosphere is to understand why changes constantly occur in the atmospheric system.

Energy in the atmosphere

Solar radiation, or insolation, occurs as short-wave radiation and is the main source of external (heat) energy input into the Earth's atmosphere system, while the Earth's motions and gravitational pull (on air masses and moisture) provides a constant internal source of energy.

The Earth's Variable Energy Profile	
Global variations	**Temporal variations**
Latitude determines the intensity of energy receipt. Insolation decreases from the equator to poles	Seasonal shifts in radiation produce large latitudinal differences in hemispherical heating between January and July (seasonal anomaly maps display this information)
Maximum solar radiation values are found in cloud-free areas such as the tropics	The distribution of heating can be influenced by the distributions of land and sea
Surplus radiation of low latitudes. Heat transferred from low to high latitudes. Deficit of radiation in high latitudes	Daily temperature changes can be related to diurnal changes/exchanges of radiation
Energy in the form of heat is transferred from the ground to the air	Reflective cloud, cooling wind and reflective snow modify daily temperatures
The unequal inputs and outputs of radiation (differences in heating and cooling) are the causes of all of our weather and climate variation	Temperature inversions may result when there is excessive loss of heat from the ground at night

KEY POINT

You must know that energy varies across the globe.

The atmosphere is reasonably 'transparent' to solar radiation, in that large amounts of energy are allowed to pass through to the ground surface. This stream of energy powers the atmosphere and the biosphere.

About 50% of the insolation received at the edge of the atmosphere is actually 'lost'. Most is scattered (by dust), reflected (by clouds) or absorbed (by clouds, dust and water vapour). Long-wave radiation emitted by terrestrial and atmospheric radiation is largely absorbed; this contrasts with solar radiation. The balance of the energy receipts is used to heat the ground and air by conduction, convection, turbulence and evaporation.

The atmosphere on the move

AQA A	U1	EDEXCEL A	A2
OCR A	U1	WJEC	A2
OCR B	U1	NICCEA	U1

Simply: uneven heating of the Earth's surface causes variations in air pressure, producing air movement or wind.

Movement in the atmosphere is probably the most obvious of its characteristics, the study of the forces that control movement being fundamental to our understanding of how energy (in the form of heat) is distributed around the globe by the global circulation.

The horizontal air movement described is called advection; vertical movements are called convection. Air motion is initiated by a pressure gradient (air density variations) between places, with movement occurring between high- and low-pressure locations.

The existence of large-scale convection models was recognised by George Hadley (in 1735) and William Ferrel (in 1889). Their postulations contributed to the equator-to-pole Convection Cell model (see diagram below).

Air does circulate in a conventional fashion, but there are consequences for the atmosphere related to this movement. They involve the rate at which the air can rise or fall (the equation of state), the ability of air to expand or contract (the thermodynamic equation) and the fact that air will continually circulate (the equation of continuity). And because we view movement from a moving platform, an apparent deflection of moving objects due to the Earth's rotation occurs, the so-called Coriolis Effect. In the mid-latitudes the pressure gradient and Coriolis are in balance. This leads to air blowing not from high to low pressure areas, but between the two, parallel to the isobars, a geostrophic wind.

Global air movements

These occur at a variety of scales. There is a close relationship between major winds and the world's pressure systems. Windbelts also contain the world's major weather systems; hurricanes in the tropics, cyclones and depressions in the mid latitudes (and at a smaller scale – tornadoes).

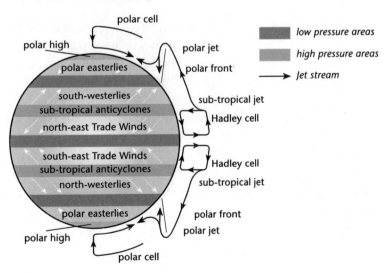

Jet streams are fast (200 km/hr) and found at about 12 000 m. They appear where warm and cold air mix. They guide weather systems (W to E) in the Northern Hemisphere. Shifts in their location can affect our weather for long periods.

Rossby waves form because of the Earth's thermal differences, the rotation of the Earth, and because of the destabilising effect of mountain ranges.

Mid-latitude movements

Found between the polar and tropical circulations is an area of complex upper air movements. The combination of the powerful Ferrel westerlies (the so-called circumpolar vortex) and the jet streams, in the mid-latitudes, affect surface weather/winds and pressure systems. These surface winds and weather belts are bound up in a series of waves known as **Rossby Waves**. The link between the Rossby waves, jet streams and weather systems in the low- and high-pressure circulation is called the **index cycle**.

Influence of the oceans on atmospheric movement

- They provide water for the hydrological cycle.
- They absorb and redistribute energy, e.g. the Gulf Stream and North Atlantic Drift benefit Western Europe by releasing heat to warm, and provide a milder climate than might otherwise be the case.
- Warmed oceans and their currents can therefore affect local and distant locations.

This phenomenon frequently appears at AS Level.

El Niño (Southern Oscillation – ENSO) 'The Christ Child'

El Niño is the unusual warming of the surface ocean layers in the Tropical Pacific areas, peaking around Christmas; it has a profound effect on world weather patterns. Occurring every few years, its cause is unclear, speculation suggests that volcanic eruptions and their release of pollutants may be a cause. Once thought to have an effect that was purely local, it's now thought the effects are felt globally. Chronologically over the last 30 years El Niño has been linked to the following:

1972 – warm water ingress, into normally cold water, along the South American coast, killed off plankton and the fishing industry.

1977 – drought in California. Severe cold on the Eastern seaboard of the USA.

1982 to 83 – catastrophic global weather. Drought in Africa and Australasia, torrential rainfall in Bolivia, Peru and Ecuador, intense winter storms on the US Pacific coast and hurricanes in Hawaii and Tahiti. At this time, locally the sea temperature rose by just 3°C! 1500 people died and there was $8 billion in property damage.

1991 to 95 – precipitation anomalies. California receives a year's rain in January 1995, cost to the USA $2 billion. Floods inundate Germany, Belgium, Holland and France. Drought affects Australia, India and South Africa.

1999 – Venezuela, Mexican and Peruvian flood disasters.

CASE STUDY

El Niño has also been linked to the spread of disease resulting from the abnormally warmed water. Cholera killed 5000 in Peru in 1991; malaria has spread to cover 45% of the world; Dengue fever, carried by mosquitoes, has since 1988 killed 4000 in the America's, mostly in Mexico; Bubonic plague increases are also linked to El Niño.

La Niña, is the opposite of El Niño. It refers to a period of cold surface water in the Pacific. The two temperatures oscillate back and forth. La Niña's effect is potentially as devastating as El Niño.

Moisture in the atmosphere

AQA A	U1	EDEXCEL A	A2
OCR A	U1	WJEC	A2
OCR B	U1	NICCEA	U1

Water is probably the most important mass moved by the air/atmosphere, with temperature defining the condition or state of the energy held. Air is mostly in its highest energy state as vapour. But, as we well know in the UK, liquid (rain) and frozen forms (snow and ice) are common.

The amount of moisture in the atmosphere is relatively small, but highly varied in time and space. When air is holding the maximum amount of water vapour possible, it is said to be saturated. The water-vapour content of the air can be expressed in terms of its absolute humidity or, more commonly, in terms of its relative humidity: the percentage ratio between the actual amount of water and the maximum amount that the air can hold at that temperature.

Evaporation and condensation are two important phase changes in the hydrological cycle and are accompanied by the absorption and liberation respectively of latent heat.

The main forms of condensation are clouds, mostly at high levels, and fogs, at or near the ground. These forms of condensation occur when air is brought to saturation point or its dew-point temperature. Condensation is assisted in the atmosphere by condensation nuclei around which liquid water can form. Condensation does not always lead to precipitation; gravity has to overcome the ability of air currents to keep the water vapour buoyant before this can occur.

Most saturation and condensation in the atmosphere take place as a result of air cooling, principally by the vertical ascent of air.

Air is forced to rise and cool by:

- orographic uplift
- frontal uplift
- large-scale convergence and ascent in low-pressure systems
- smaller-scale convective currents.

Stability of the air – Lapse rates

Advectional processes continually seek to establish a stable atmosphere, in terms of the distribution of energy. Air masses with their uniform temperature and humidity typify this stability. However, within air-masses parcels/pockets/balloons or bubbles of air must move vertically (or adiabatically) up or down, especially when air masses move over irregular relief or variably heated land masses. The stability or otherwise of these smaller bubbles of air has dramatic effects in terms of local weather.

In general, temperature falls as you move higher into the atmosphere, this is known as the ELR (**Environmental Lapse Rate**). As pockets of rising air gain height they cool, expansion occurs, further energy (kinetic) is lost and the pocket of air cools still further. So long as it is not saturated it cools at a fixed rate, the DALR (**Dry Adiabatic Lapse Rate**). Once the rising air reaches the condensation level it becomes saturated. Latent heat is produced and water vapour turns to water. Air cooling is now reduced, and the air now cools at a slower rate than previously. This rate is known as the SALR (**Saturated Adiabatic Lapse Rate**).

The relationship between the ELR and both the DALR and the SALR influences the temperature, and thus the density and buoyancy of a vertically displaced parcel of air, in comparison with its atmospheric surroundings. It thus determines the character of air stability.

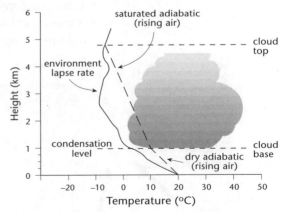

Unstable air occurs when a vertically displaced warm air parcel is encouraged to rise, at sufficient height it cools and condenses into clouds. In this case, the ELR is greater than the DALR or SALR.

This will happen when the air is cooler than surrounding air.

With **stable air**, a parcel of air displaced vertically upwards or downwards in the atmosphere will tend to return to its original position. Here the ELR is less than the DALR or SALR.

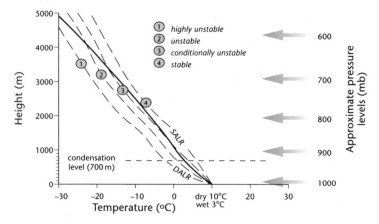

Air stability is an important concept as it determines the buoyancy of the air and thus development of cloud and fog. Cumulus clouds, often of large vertical extent, are characteristic of unstable air, whereas horizontally developed stratus clouds tend to form in more stable air.

3.2 Weather systems

After studying this section you should understand:

- *the mid-latitude circulation, air masses and fronts*
- *the causes and effects of extreme weather*
- *how man influences and affects the atmosphere*

LEARNING SUMMARY

Britain and Western Europe's weather and climate

AQA A	U1	EDEXCEL A	A2
OCR A	U1	WJEC	A2
OCR B	U1	NICCEA	U1

Britain has a Cool Temperate Humid (Maritime) Climate. Features of this climate are:

- it is typically found in the mid-latitudes 40° and 60°N (or S)
- marine influences are predominant, and can influence areas far inland
- eastward-moving air bringing depressions (cyclonic conditions) with characteristics of temperature and humidity derived from the oceans
- the weather can be extremely changeable and variable (e.g. continental high pressure extends at times causing severe cold and within hours southwesterly flow is re-established and warm damp conditions reappear)
- winters are mild, the oceans effect is marked
- summers are warm, 13° to 18°C is common in July (the oceans can cool summer temperatures too)

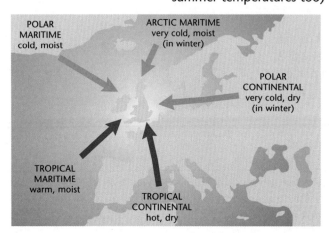

- drought conditions are rare; but the ingress of North African air can bring spells of prolonged hot weather
- rain occurs throughout the year; but varies from place to place based on the relief (it chiefly runs in on cyclonic fronts and decreases to the east)
- depressions (cyclonic conditions, lows) are less common in summer than in winter, and generally bring warm but wet conditions
- anticyclones bring hot, sunny weather in summer and fog, snow, cold and still conditions in winter.

Local weather processes and patterns

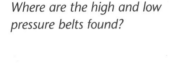

Low pressure	High pressure
What is it?	*What is it?*
If an area has pressure that is lower than its surroundings, then it is termed low pressure. Cyclones and depressions are low-pressure systems.	If an area has pressure that is greater than its surroundings, then it is termed a high-pressure area. Anticyclones are high-pressure systems.

Where are the high and low pressure belts found?

The weather processes and patterns that affect the UK need to be understood and learned for AS

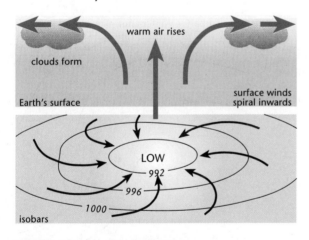

North Pole

HIGH

LOW

(60°N) — — — — — — — — Tropic of Cancer

HIGH

LOW — Equator

HIGH — Tropic of Capricorn

(60°S)

LOW

HIGH

South Pole

Causes of low pressure

Causes of high pressure

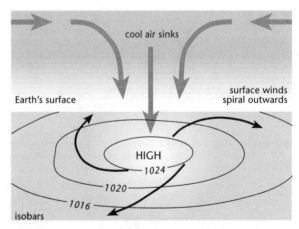

What are the weather effects?

Unstable rising air causes unsettled and changeable weather.

Typically depressions bring:

- cloudy skies
- low levels of sunshine
- wet weather
- temperatures that are mild for the time of year
- windy conditions
- changeable weather.

What are the weather effects?

Typically these pressure belts bring stable weather conditions:

- clear skies
- sunshine
- dry weather
- high day and low night temperatures
- calm weather
- dew and frost
- fog and mist
- thunderstorms
- snow in winter.

Low pressure
How fronts form and the typical passage of weather

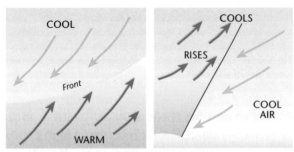

Stage 1: Warm and cold air meet – warm air rises

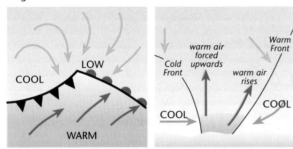

Stage 2: Low pressure develops

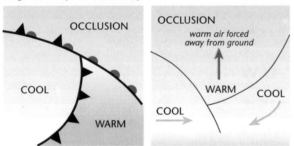

Stage 3: Fully developed frontal system formed as warm air rises

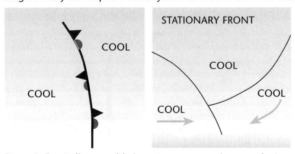

Stage 4: Front dies as cold air squeezes warm air upwards

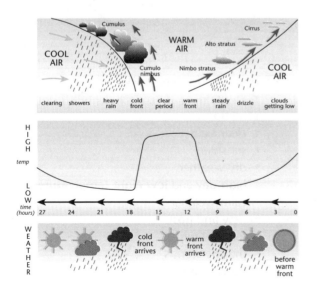

High pressure
The 'synoptic' winter and summer anticyclone

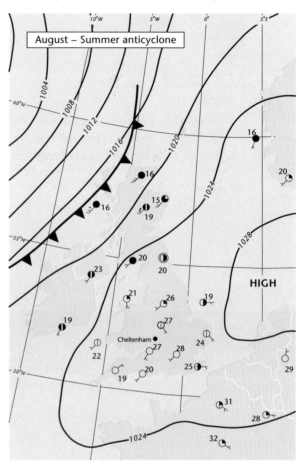

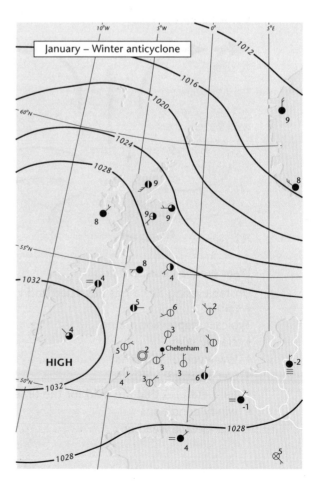

The low pressure synoptic chart for April over the UK

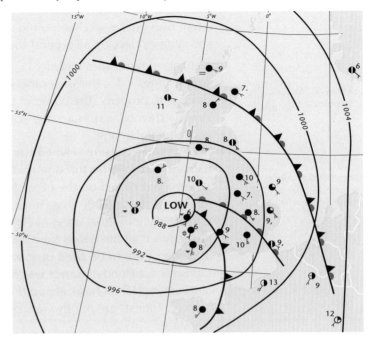

Blocking anticyclone

This tends to be a large circulation that is no longer in the mid-latitudinal frontal belt. It effectively blocks the west to east passage of air, usually for several weeks. Within the high, typical and persistent anti-cyclonic conditions (for winter or summer, depending when the block occurs) prevail.

Severe weather and related phenomena

AQA A	U1	EDEXCEL A	A2
OCR A	U1	EDEXCEL B	A2
OCR B	U1	WJEC	A2
		NICCEA	U1

Extremes of precipitation and temperature cause most of our weather hazards. Extremes of air movement surprisingly create more minor effects.

Tropical cyclones

These are classified differently around the world, cyclones hit Southern Asia (between June and November), the Willy Willies hit Australia (between January and March), and typhoons hit the western Pacific (between July and October). Typically these generate winds of between 40 and 119 km/hour. Hurricanes are also a form of tropical cyclone. The intensity of the hurricane and the extreme effects they have had on the Gulf coast of the eastern seaboard of the USA ensure that this form of cyclone has been well researched and documented.

18% of the world's population live on coastal areas that are under threat from tropical cyclones. On average 20000 people/year lose their lives to tropical cyclones.

Wind spins around an eye of 25 to 50 km this decreases in size as hurricanes speed up.

The ITCZ is where convergent air meets near the equator. This area is intermittent and can vary its position.

Hurricanes

Certain atmospheric and ocean conditions seem to be at the root of their formation. The following develops this 'formation' theme.

- Generally these intense low-pressure systems (less than 920 Mb is common) form between 30° N and S, in an area called the **Intertropical Convergence Zone**. With Coriolis Effect at its greatest in the tropics the systems' rotation can be easily initiated (winds in the order of 119 to 300 km/hr are common).
- The sea in this area is well heated, and exceeds the minimum 26° to 27°C necessary (between August and October) for latent heat to be released (because of condensation of water vapour) to strengthen the hurricane (relative humidity of 60% is common). The ocean then provides the initiating and sustaining energy for hurricanes. They quickly 'die' once they move onto the land. The high levels of moisture held also point to a sea origin (typically in the order of 10 to 25 cm/day).

- Towering (15 km is common) cumulo-nimbus clouds form around the central eye in highly unstable conditions. These clouds further fuel the hurricane as latent heat energy exchanges moisture from gas to liquid. Spinning weather sub-systems develop all around the main hurricane mass.

The effects of hurricanes include:

- storm surges – it is this phenomenon that potentially is the biggest killer
- damage to property (the power of the hurricane is such it turns property instantly into matchwood. e.g. Hurricane Andrew, in Florida, left 250 000 homeless)
- agricultural damage
- huge insurance claims – when Hurricane Andrew struck in 1992, 25% of damage was paid for through insurance (some $5.6 billion), the federal government picked up the other $19 billion bill!
- refugees – the 250 000 that Hurricane Andrew left homeless moved temporarily into other areas of Florida and into neighbouring states
- property – if all the debris from the properties that hurricane Andrew destroyed were to be piled up it would have towered 300 stories high!
- tourism – for Florida, hurricanes create a unique situation. It is a major tourist destination. Florida has suffered three of the five deadliest hits in the USA this century. Tourists are rightly wary of holidaying during the hurricane season in this state.

> Hurricanes distribute energy polewards.

> Between 1900 and 1991 there were 152 direct hurricane hits on the USA.

> Examiners look for a good mix of process and factual understanding. Simple diagrams like this convey vast amounts of information extremely rapidly.

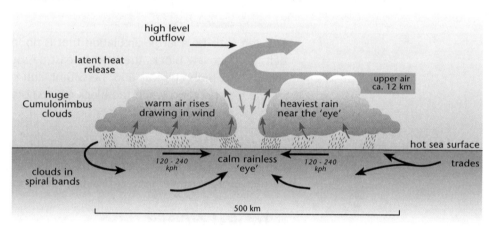

> Newspapers are a good source of up-to-date information. AS questions frequently use this topic information.

Very occasionally hurricanes are re-vitalised in the mid-latitudes and therefore can reach the shores of Britain, though in an extremely modified form.

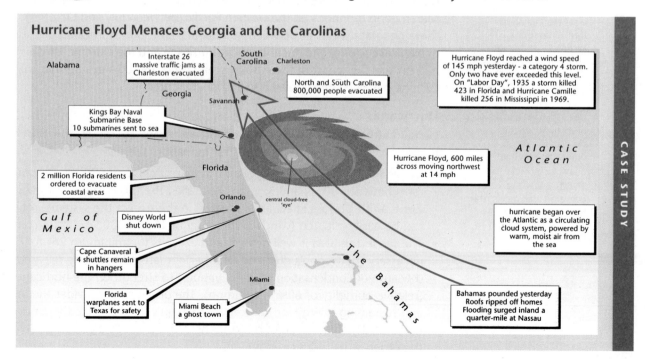

Tornadoes

These are a concentration of cyclonically spinning air (about 160 km across), found overland, rather than over water. Most often a visible cloud forms from a large cumulo-nimbus origin. The tremendous rotations are initiated by rapid convergence at the base of the cumulo-nimbus cloud, as rapid updraughts develop. Tornadoes are common over the mid-west and southern states of the USA during the heat of early summer afternoons. Tornadoes are an increasingly common and observed feature in the UK, e.g. The Selly Oak twister of summer 1999, but perhaps more dramatic and bizarre was the Selsey Tornado, of 1999. N.B. Tornadoes may be a by-product of hurricane development.

The SE Asian Monsoon

The word monsoon means season.

The SE Asian monsoon is the major disturber of the global atmospheric circulation. It displaces masses of energy to the North and South between 60° and 180° E.

Causes

1 **Pressure and wind**
   The displacement of pressure and wind is caused by the direct heating of the Earth, this causes the land to heat up disproportionately to the sea. The Earth also cools disproportionately to the sea.

2 **The ITCZ/ITD (inter tropical discontinuity)**
   Seasonal shifts of the ITD and its windy belts.

3 **Mountain influence**
   Without the mountains, patterns of rainfall would be different in India.

4 **Concentrations of CO_2**
   Reduced CO_2 over India, because the effect of uplift over the Himalayas causes less heat to be absorbed, which leads to cooling.

5 **Jet-stream positions**
   In India the seasonal changes brought about by the above cause two markedly different monsoon periods.

The winter monsoon

Westerly jets split around the Tibetan Plateau. Descending air causes high pressure over central Asia resulting in out-blowing northeast winds across Asia, clear skies, and little rain. Sunny weather results.

The summer monsoon

Beginning in the spring and finishing about June. The phases of the summer monsoon are:

- in March the northern jet dominates the area
- the sun is directly over India
- the ITD moves to an area south of the Himalayas
- low pressure moves northwards – the so-called monsoon trough
- one part of the trough moves west over India, the other to the east over the Bay of Bengal, they meet and join over northern India in June
- low pressure continues to deepen and develop over the area and the North Indian Ocean. Under the intense isolation of the area's clear skies, warm moist air starts the cycle of the so-called monsoon rains
- lots of warm moist air is drawn in over India from the North Indian Ocean: intense rainfall results.

The monsoon brings drought-breaking rainfall and wind, and invariably initiates devastating subtropical cyclones (e.g. Orissa, India 1999).

Impacts

Agricultural yields increase during the 'good' monsoon years. This has to be balanced against flooding and loss of life.

CASE STUDY

55

Weather: human aspects

AQA A	U1	EDEXCEL A	A2
AQA B	UB	EDEXCEL B	A2
OCR A	U1	WJEC	A2
OCR B	U1	NICCEA	U1

Every aspect of our lives is affected by weather; the table below summarises the effects on various areas of economic and social activity.

Effects on the economy

> Know and understand the effects weather has on us!

Agriculture	Animals and livestock get distressed in excessive heat or cold, particularly young animals. Fodder and grazing need to be available at all time for animals, this is difficult in winter. Crops. Few grow above 60° N or S. Most crops can be affected by variations in day-to-day weather fluctuations. A late frost affects the coffee crop. Droughts are threat to crops. Pests tend to be weather dependent (locusts).
Fishing	Is completely weather dependent. Icing of the superstructure leads to disasters, as do storm conditions.
Construction	Concrete laying and groundwork cannot take place in extreme cold or wet periods. Tenders are often adjusted to allow for weather risks. Using the services of the weather forecaster saves £120 million/yr for the construction industry in the UK.
Transport	Bad/adverse weather causes delays and accidents. In the UK airlines can save millions of pounds per year, local highway divisions can save by only gritting where necessary, and shipping can save fuel and time, all by using forecasting to help them be more efficient.
Power	There is a strong and obvious link between weather and fuel used. Most utilities estimate, using past weather records, when greatest demand will be.
Television	Reception is affected by rapid changes in temperature and humidity. Anticyclonic conditions in particular affect reception.
Business and Retail	Most businesses would be unwise to ignore the weather. E.g. in summer, supermarkets stock up on BBQ fuels and foods, salad products and fruit. Cinemas lose their attraction. DIY enthusiasts buy equipment and plants. Ice cream sellers do a roaring trade. In winter, stocks of antifreeze and batteries for cars sell quickly. Plumbers attend more leaks and bursts. Pharmacists dispense more medicine. Wet weather equates to more business for retail outlets in tourist areas, and so on.
Leisure and Sport	By using forecasting we can plan our activities for maximum enjoyment and safety. In the UK indoor parks (Centre Parcs) and covered stadia (Millennium Stadium in Wales / the Millennium Dome and the new Wembley Stadium Complex) are all responses, perhaps, to our unreliable weather.
Health	Housing, dress and way of life all relate to the weather and climate of our respective home area. Our comfort (addressed by air conditioning or heating), activity (i.e. cold restricts activity) and health (asthma, heat-stroke, hypothermia and hayfever) are all affected.

KEY POINT

Threats to our ecosystems

The ozone layer

> Too much UV can cause cataracts and crop yield drop, and plankton are damaged. The latter not only save food chains, but also have a regulatory exchange effect on CO_2. Plankton can regulate global temperatures. They can affect global warming! (See coastal chapter.)

Ozone is a thin layer found in the stratosphere, 10 km to 50 km above the ground. Being very unstable, ozone is easily broken down by high-energy UV light; the process of breakdown causes absorption. This blocking effect is

Chlorine is released readily in the cold (below –80°C) sunless skies of the Poles.

A 1% decrease in ozone causes 70 000 new skin cancers and 100 000 new cataracts.

N.B. Ozone depletion was first noted in 1975 by the British Antarctic Survey.

• We add 10 tonnes/person/year of CO_2 to the atmosphere.
• Current values = 240 ppm, during the last ice age there were 200 ppm

N.B. Our memories are short, in the 1970s we were warned about a return of the ice to the land! During the periods between the last four ice ages temperatures were 5°C warmer than today!

beneficial to us because high intensity UV can cause skin cancer.

Ozone is being depleted by mans use of CFCs (chloroflourocarbons) in industry, as a coolant in refrigerators and in aerosol propellants. Once released, CFCs rise high in the atmosphere, releasing chlorine. Chlorine is the destroyer of ozone.

By April 1992 UK and European scientists confirmed the American Upper Atmosphere Research Institute's fears that chlorine levels were high, and rising, and that ozone was being destroyed more rapidly than ever before. If levels of ozone could be reduced or frozen at 1986 levels, as in the Montreal Protocol, then ozone levels might recover by the end of this century! What seems certain is that ozone levels will certainly deteriorate further before they improve.

Smog

The smogs we experience today are vastly different from those of the 1950s. London, for instance, was known as 'Old Smokey' or the 'Smoke' because of the thick smoky fogs that would often envelop the city. These only disappeared in the early 1960s when legislation and smokeless fuels ended the polluting effects of bituminous coal. Today we have to deal with photochemical smog, the more lethal relation of smog. Photochemical smog, produced by the exhaust gases of vehicles and industry, in still, warm, clear sunny conditions contributes to a build up of poisonous ground level ozone. Small amounts at ground level are lethal, affecting breathing, causing conjunctivitis-type irritations and affecting plant and animal tissue.

The greenhouse effect and global warming

Warming near the Earth's surface results in the atmosphere trapping the Sun's heat: without the greenhouse effect the Earth would be 33°C cooler. CO_2 is a major contributor to the greenhouse effect and with water vapour it absorbs vast amounts of heat. Since the Industrial Revolution we have dumped vast amounts of CO_2 into the atmosphere. Once emitted, a CO_2 molecule stays active in the atmosphere for about 200 years. Some computer models estimate a 1.5°C to 4.5°C increase in temperature over the next two centuries because of global warming. CO_2 from human activity is therefore a major cause of global warming, along with methane from termite colonies, flatulent cattle and rice paddy fields.

It should be remembered that global warming has been widely debated, that climate varies naturally and that in fact actual data that really does support the theory is difficult to come by and verify. Certainly some areas in the world have been hotter and drier and river flows have varied. However, many people believe that the data suggest a new ice age is upon us!

Best scientific evidence supports a significant increase in the Earth's temperature over the last 100 years (this is well within normal averages/fluctuations of the Earth's atmosphere). This increase will affect sea level; the Marshall Islands, Maldives and Kiribati islands, lower areas of the Nile and Bengal deltas could all be threatened by increases in sea level.

In the developed world we are attempting to reduce levels of CO_2, for instance, by using energy efficient lights we might reduce CO_2, sulphur dioxide and nitrogen dioxide emissions. Burning leaner and cleaner fuels and the implementation of energy efficiency campaigns will reduce the amounts of greenhouse gases released into the atmosphere.

Acid rain

Quantities of sulphur dioxide and oxides of nitrogen are emitted into the atmosphere from industrial activity and vehicles. In the atmosphere chemical changes turn rain into weak acids (in the order of pH 4.3). *Why is it a problem?* In the late 1960s lakes became acidic and fish populations plummeted in Scandinavia, caused by pollution from Western Europe. Germany has been particularly affected, in the mid-80s 58% of trees showed some signs of damage. Over the last twenty years Germany has spent £300 million, attempting to sort out its problem forests. Acid rain also 'rots' buildings, the Houses of Parliament being a prime example. *The future?* Various Clean Air Acts in the Westernised countries have reduced the problem significantly. Remedially, lakes can be limed to neutralise acids.

Aspects of local climate

Air flow in mountains (anabatic and katabatic winds)

During the day when conditions are calm, warm air blows up the valley in response to the heating of the air in contact with the valley slopes. At night the reverse happens. A return wind at night completes the circulation. Accumulations of cold air in valley bottoms can cause frost hollows to form.

Land and sea breezes

On warm days air over the land heats up and expands. This tilts the pressure gradient. The result is that a small landward blowing circulatory system develops, with compensating outblowing wind aloft. The reverse happens at night. These breezes have a marked effect on coastal climates.

Mountain climates

Few mountain climates fit into any climate scheme easily. For instance, in tropical areas increases in height are the equivalent of an increase in latitude. In Africa and other equatorial continents and countries, mountains are important for their effect on precipitation patterns and for the moderating effect they have on temperatures.

Urban climates

Buildings interfere with both wind and airflow patterns and they change temperature regimes. In the high-rise cities of the world, wind is channelled down streets, the so-called **venturi effect**. These effects cause massive turbulence in city streets, particularly at night and especially in winter. Cities tend to have lower humidity rates, because of a lack of vegetation. Evaporation tends to be high, combined with turbulence; thunderstorms over or near cities are common. Increased light precipitation over and around cities is an observable feature, and may contrast markedly to the surrounding countryside.

Heat island effect

The contribution of industry, central heating, the heat retaining and emitting properties of the fabric of the city, and the blanket of pollution, cause temperatures recorded over the city to be one or two degrees higher than the surrounding countryside. If plotted on an isotherm map, temperatures show a marked decline from the central area to the edge of the city.

Sample question and model answer

1

The diagram below shows the environmental lapse rate (ELR) for a vertical section of the atmosphere.

The dry adiabatic lapse rate (DALR) is 1°C per 100 metres, and you should assume that the saturated adiabatic lapse rate (SALR) is 0.5°C per 100 metres.

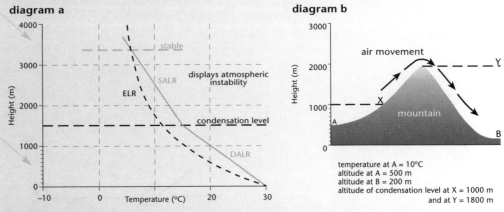

(a) On the diagram **a**:
  (i) Draw lines for the DALR and the SALR for an air 'parcel' rising from ground level where the air temperature is 30°C.

  (Answer in orange on diagram)

  (ii) Indicate the height at which the air 'parcel' becomes stable.

(b) (i) What is meant by the term 'adiabatic'?   [2]

   The process of temperature change as a result of pressure change, there is no exchange of heat or energy with the environment.

  (ii) Explain why there is a difference between the dry adiabatic lapse rate and the saturated adiabatic lapse rate.   [3]

   In unsaturated air, cooling occurs, because of expansion at a consistent 10°C/1000m. Once saturated it cools more slowly.

(c) Diagram **b** shows air movement across a mountain.

  (i) What is the temperature difference between locations A and B?   [2]

   + 17°C

  (ii) Explain this temperature difference.

   Rising air cools at the DALR, up to the dew point. It cools at the SALR until the summit is reached. Descending air warms, adiabatically increasing temperatures locally – the Föhn effect.

  (iii) Comment on two effects, other than temperature, that the air movement has on weather conditions in the area around B.   [4]

   1. Because leeward winds have dried out the air adiabatically, the winds are able to evaporate atmospheric moisture easily.
   2. Very little rain falls.

(d) Relate adiabatic processes to cloud formation?   [5]

   The bubble of rising air cools. At dew point it becomes saturated and condensation sets in. This is the cloud base level. The air continues to cool, but more slowly now. It starts to lose its buoyancy and the top of the cloud forms.

   Edexcel (modified)

Practice examination questions

Section A

The diagram below shows the position of areas of high and low pressure and the direction of the surface winds blowing between them.

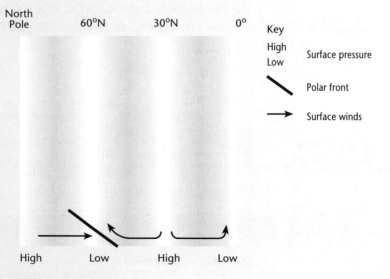

1 Complete the above diagram.

   (a) Indicate a Hadley Cell.   [2]

   (b) Account for the size, shape and directions of air movements drawn.   [2]

   (c) Outline the differences between the area of low pressure around the
        Equator and those at the polar front using the following **three** headings:

        − reasons for the air rising;
        − type of cloud that usually forms;
        − precipitation characteristics.   [6]

   (d) How can El Niño affect the weather and economics of the tropical and
        subtropical zones?   [3]

   (e) Explain how the weather associated with periods of long winter cold affect
        humans.   [8]

Section B

2 (a) What are the consequences of global warming; and how will they effect
        our economic and social well being?   [10]

   (b) Using one country as a focus, outline the impacts of global warming
        upon it.   [10]

AQA

Chapter 4

Earth challenge

The following topics are covered in this chapter:

- The lithosphere and tectonic processes
- Rock type and weathering processes

4.1 The lithosphere and tectonic processes ✓

After studying this section you should understand:

- the contribution continental drift and plate tectonics theory offers to our understanding of volcanic and earthquake activity
- that volcanoes and earthquakes are an ever-present danger
- that tectonic activity creates as well as destroys
- that man can increasingly predict and prepare for volcanic and earthquake activity
- that differences exist in human response to tectonic hazards based on levels of economic development

LEARNING SUMMARY

Isostacy: is the principle of flotation or buoyancy. Differences in density between continental and oceanic areas explain why continents stand high above oceanic basins.

The Earth has been in existence for approximately 4.6 billion years. During that time the character, position and distribution of the land and sea has changed many times. However, we have a pretty good understanding of the internal structure of the Earth from earthquakes, volcanoes and plate tectonic activity.

The internal structure of the Earth

NIFE – Alloys of Nickel and Iron at 4000°C. The outer core is liquid and spins with the Earth. Thought to be the source of Earth's magnetism

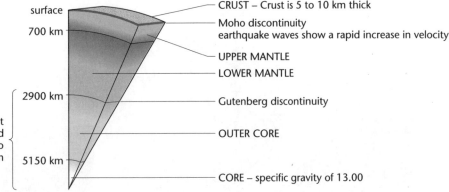

- CRUST – Crust is 5 to 10 km thick
- Moho discontinuity earthquake waves show a rapid increase in velocity
- UPPER MANTLE
- LOWER MANTLE
- Gutenberg discontinuity
- OUTER CORE
- CORE – specific gravity of 13.00

surface
700 km
2900 km
5150 km

Continental drift and plate tectonics

AQA A	A2	EDEXCEL A	U1
AQA B	A2	EDEXCEL B	A2
OCR A	U1	WJEC	U1
OCR B	A2	NICCEA	A2

Many theories are comparatively recent!

Alfred Wegener was the first to produce the theory of continental drift in his book *The origin of continents and oceans*. He noted the jigsaw fit of the continents, the similarity of the rock structures and fossils on either side of the Atlantic, and the fact that there is abundant coal in Antarctica. Arthur Holmes and Alexander Du Toit kept the hypothesis alive. They envisaged rafts of lighter granitic continental rocks (SIAL – Si+Al) floating on denser basaltic oceanic rocks (SIMA – Si + Ma).

It was only when good maps of the ocean floor were produced showing underwater mountains, ridges, island arcs, trenches and plains, that scientists became aware of plate structures. Plate tectonics have proved a unifying theory – widely accepted by the late 1960s.

The Earth's outer layer is made up of granitic, irregular-shaped plates. There are eight major ones and a dozen or so smaller ones. The plates are about 50 to 70 km in depth, rigid and are part crust and part upper mantle (**lithosphere** = the solid outer rind). They ride on a less rigid layer, the **asthenosphere** where the rock is almost at its melting point.

There is a strong link between volcanoes, earthquakes and plate boundaries.

Most of the world's great mountain ranges, most destructive earthquakes and volcanoes occur at the boundaries (e.g. the Pacific Ring of Fire). There are three types of boundary: accreting, consuming and conservative. While most of the present day ocean floor is less than 200 million years old, some continental rocks have an estimated age of 4000 million years.

Consuming boundaries

(Alternatives = destructive/collision and subduction zones.)

Plates move together and material is destroyed. One plate rides over the top of another pushing it back deep into the Earth to be remelted. Some plates are rather like rocky conveyor belts. The continents are too light to be pushed under. When continents carried on two separate plates collide, the ground buckles and folds and a new mountain chain is born. Two distinctive types of consuming boundary occur: subduction zones, with ocean trenches and island arcs, and collision zones with mountain ranges.

Labelling a simple diagram is common at AS Level.

Ocean trenches are typically 1600 km or more in length, 96 km wide and 2 to 3 km below the surrounding ocean and are V-shaped. The deepest soundings have been made in the Marianas Trench at 11 040 m. The deepest in the Atlantic is the Puerto Rica Trench at 8648 m. The descending plate dips at an angle of about 45°. It takes about 10 million years to reach a depth of 720 km; at this point it loses its rigid characteristics and has heated up sufficiently to blend in with the surrounding mantle.

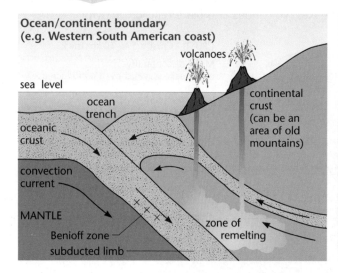

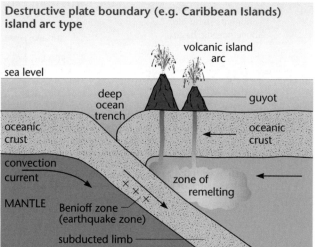

Volcanic Island Arcs are situated on the over-riding plate 100 to 200 km behind the trench. Two-thirds of the 800 recorded volcanoes during recorded history are associated with island arcs. The Peru–Chile trench is near the coast but if it were further out to sea, the volcanoes of the Andes would also be further out to sea. Evidence suggests that island arcs have played an important part in shaping the face of the Earth by contributing much of the material that makes up the continents.

Most volcanoes and destructive earthquakes are associated with consuming boundaries.

Accreting boundaries

(Alternatives = constructive/spreading ridges.)

The plates move apart, molten material and hot rock seep to the surface from the mantle to cool and form new lithosphere. Almost all these boundaries lie deep beneath the oceans where they form mid-ocean ridges (one exception is Iceland). There is a continuous system twisting and turning for 75 000 km around the world.

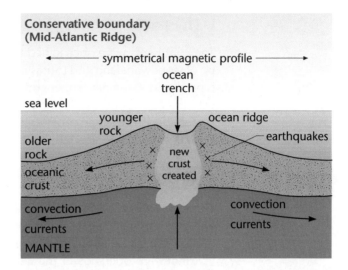

Sir Edward Bullard measured the heat coming to the surface along the mid-oceanic ridges and found that it was significantly higher than elsewhere along the sea bottom. There are many deep cracks, and a rift valley with an average width of 32 km and up to 1525 m beneath the crest runs down the centre of the range. Most tremors take place in or near it.

Palaeomagnetic studies

Frequent questions ask for a knowledge of the evidence for continental drift and plate tectonics.

The Earth is a weak magnet. The magnetic field behaves strangely. The magnetic poles wander slowly about and from time to time the magnetic field reverses itself (171 times in the last 76 million years). There is a record, frozen in rocks containing iron oxides. Lava shows no magnetism, but as the rock cools it becomes magnetised in the direction of the north magnetic pole and then retains this record throughout time. Magnetic readings from the ocean floor form a zebra-striped pattern proving sea-floor spreading. ('Sea-floor spreading' is attributed to the work of Vine and Matthew). Spreading rates vary from 2 to 20 cm per year. The distance between Europe and North America has only increased by about 15 m in nearly 500 years.

Conservative boundaries

(Alternative = transform boundary.)

No material is added or destroyed. Two plates grind and slip past each other in a series of jerks, e.g. San Andreas Fault, where perhaps in 10 million years Los Angeles will be alongside San Francisco. Transform faults occur widely in the fracture zones associated with spreading ridges. Pressure, rotation and contraction cause them.

The mechanism

The force that drives the plates is to a certain extent still a mystery but their movement certainly suggests convection currents in the mantle. In 1971, J.W. Morgan developed the Hot Spot or Plume Theory. He suggested that giant pipes that carry up hot material from deep inside the Earth to the surface powered the plates. These plumes are up to 100 km in diameter. 99% of the material spreads out beneath the plate in a circular pattern like a thunderhead cloud. 1% reaches the surface. Where several hot spots punch the surface, the lithosphere may break. This would form a crack and lava could seep to the surface to form an accreting boundary. He located 20 possible hot spots – 7 main ones beneath the Mid-Atlantic ridge and 4 along the East Pacific Rise. They may not flow continuously. The Hawaiian Islands could have been formed by this method. Most rock of the spreading ridges is basalt from the upper asthenosphere. The rock that forms Iceland is from much deeper – compatible with the hot spot theory.

The pattern of change

The continents were once joined to form one landmass; Pangaea (all lands), made up of Gondwanaland and Laurasia. It began splitting about 200 million years ago. It is predicted that a new super-continent will form again 200 million years from now with the centre of the old Pangaea becoming the coastlines of the new landmass. It is probable that a single continent existed 700 million years ago before breaking up. By 570 million years ago four continents were moving about the globe.

Volcanoes

AQA A	A2	EDEXCEL A	U1
AQA B	A2	EDEXCEL B	A2
OCR A	U1	WJEC	U1
OCR B	A2	NICCEA	A2

These play a key role in shaping the planet. They probably created our atmosphere and they certainly release water vapour onto the surface of the planet, creating the oceans. Every year about 50 or so of the Earth's 700 active volcanoes erupt. Simply, volcanoes are openings in the Earth's crust through which magma, molten rock and ash erupt onto the land.

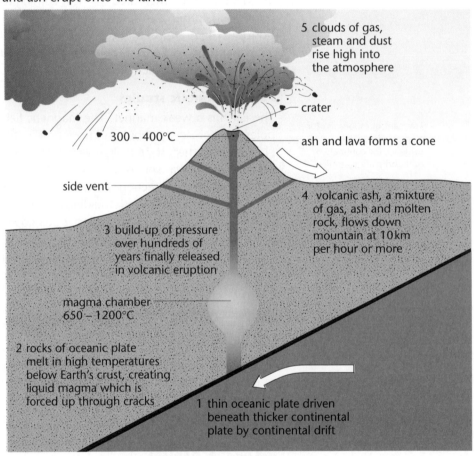

5 clouds of gas, steam and dust rise high into the atmosphere

crater

300 – 400°C

ash and lava forms a cone

side vent

4 volcanic ash, a mixture of gas, ash and molten rock, flows down mountain at 10 km per hour or more

3 build-up of pressure over hundreds of years finally released in volcanic eruption

magma chamber 650 – 1200°C

2 rocks of oceanic plate melt in high temperatures below Earth's crust, creating liquid magma which is forced up through cracks

1 thin oceanic plate driven beneath thicker continental plate by continental drift

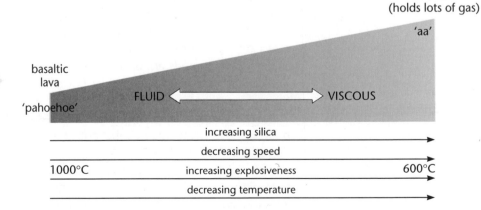

acid lava (holds lots of gas)

'aa'

basaltic lava

'pahoehoe'

FLUID ⟵⟶ VISCOUS

increasing silica

decreasing speed

1000°C   increasing explosiveness   600°C

decreasing temperature

Simple and effective diagram. Such diagrams should appear in your essays, for instance.

A knowledge of these examples of volcanic hazard is important at AS Level along with appropriate case studies.

The hazards resulting from volcanic activity are divided thus:

Hazards	Details	Examples
Lava flows	Two main types are *aa* and *pahoehoe*. When hot enough it flows like water.	The 3337 m Mount Etna on Sicily is a frequent eruptor; basalt lava from it runs rapidly down its flanks.
Lahars	These are torrents of mud, moving at 90 km/hr. There are two main reasons for their development: • increased rain because of volcanic activity washes soot and ash downhill • heat from the eruption melts snow and ice, which picks up debris.	Armero, at the head of the Langunilla Valley in Colombia was buried under a mudflow 20+m thick in November 1985.
Ballistic and tephra clouds	This is the solid material such as dust, ash, and cinders that is thrown from the volcano.	During the eruptions of Mount Pinatubo in the Philippines in 1991 rocks the size of tennis balls were hurled 50 km away!
Gases and acid rain	Gases released encompass a wide range of chemicals, from CO_2 to hydrogen sulphide.	In August 1986 a huge volume of CO_2 escaped from Lake Nyos in Cameroon, West Africa.
Nuée ardentee	A mixture of superheated rock and gas.	Mount St. Helens, May 1980. But more dramatic was the eruption of Mount Pelee, May 1902, on Martinique.
Jokulhlaups	Sudden outbursts of water due to a build up of sub-glacial meltwater due to volcanic activity.	Caused by the meltdown of ice by the volcano Loki under Vatnajokull ice-sheet.
Tsunamis	Form when water is vertically displaced. Caused by volcanic eruptions (and earthquakes). Tsunamis move at 800 km/hr with a height near land of 20 m or more.	Noteworthy here is the eruption of Krakatoa in 1883.

KEY POINT

Other dangers include landslides, fire, disease, and volcanic winters (where atmospheric ash blocks out the sun, and crop yields drop).

Volcanoes on the whole display themselves typically as a conical shape, though this depends on the nature of the material and type of eruption. They are classified in a number of ways, i.e. type of flow, type of eruption and level of activity.

Volcanic predictability

At present 20% of the worlds volcanoes are watched 24hr/day. Most are in MEDCs.

Scientists are becoming increasingly proficient at predicting volcanic eruptions. The main methods of prediction include assessing the pre-cursors to eruption.

- Seismometers measure earthquake activity which increases near a volcanic eruption.
- Gas levels – sulphur emissions increase near an eruption.
- Lasers/tiltmeters detect changes in the slope of the volcano and assess swelling of the cone.
- Ultrasound is increasingly utilised to monitor changes.
- Thermal anomalies are also measured to assess activity.

People continue to live in volcanically active areas because:

- lava weathers to give fertile soils, e.g. Indonesia, the West Indies
- precious stones and minerals form in such areas, e.g. South Africa
- superheated water can be used for geothermal power, e.g. Iceland
- areas are tourist attractors, e.g. Tenerife and Lanzarote
- products of vulcanicity can be used for building, e.g. ignimbrite blocks.

Responding to the volcano hazard

Little can be done to control volcanoes. Only lava has been successfully dealt with, e.g. Eldafells lavaflow on Heimaey, Iceland was halted in 1973 with water sprays and Etna's lava by damming and banking in 1983. Barriers and conduits have been built in Indonesia to guide lava and lahars away from vulnerable areas. Steep and strengthened ash-shedding roofs have been built in the Philippines. In other areas like Montserrat people physically move their house!

At best a 'volcanic' community can be prepared for disaster. Through accurate land-use planning, hazard assessment and mapping, and through utilising overseas technical know-how and aid resources before, during and after events, many of the effects can be negated.

Earthquakes

AQA A	A2	EDEXCEL A	U1
AQA B	A2	EDEXCEL B	A2
OCR A	U1	WJEC	A2
OCR B	A2	NICCEA	U1

The enormous quantities of energy required to set the ground in motion cannot be maintained, the effect is rapidly over, and damage tends to be confined to small areas.

Over the globe there are probably 500 000 earthquakes/year of which a hundred or so will do damage. They account for about 10% of natural disaster deaths per year.

Earthquakes are basically shock waves that are transmitted from a **focus**, which can lie anywhere from the surface to 700 km beneath the Earth's crust. The most damaging quakes have foci that are close to the surface and tend to arise along active plate boundaries. Earthquakes result from sudden movements along geological faults, bumping and grinding against each other in response to convection currents in the deep mantle rocks of the crust (or because of volcanic eruptions). Earthquakes generate a variety of shock waves, from the above movements, that have differing effects. Three types of seismic wave have been identified – P (primary or push) waves, S (secondary or shear) waves and surface (or long) waves (see below). The speed of these waves varies according to the properties of the rocks through which they pass (P waves are the fastest and surface waves the slowest). P waves are compressional in nature and tend to cause the low rumbling associated with quakes and do damage to the area around the **epicentre**. They tend not to be as damaging as the shear and surface waves. Shear waves cause the first shaking motion, vibration occurring at right angles to the direction of travel. Particle motion in surface waves is greatest at the surface and dies out at depth. Surface waves have the potential to shake large areas of the globe.

Magnitude

The magnitude and intensity of earthquakes depends on:

- the depth of shock origin
- the nature of the surface layer, e.g. soft materials amplify the shock
- the nature of the overlying material – hard rock absorbs/soft amplifies
- the numbers of people and buildings over the focus.

Earthquake magnitude is measured on the Richter Scale, using a seismometer. Each unit on the Richter Scale represents a x10 increase in wave amplitude. Intensity is measured on the Mercalli Scale.

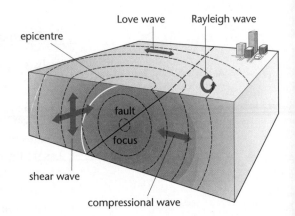

Effects

Earthquakes can cause:

- vertical or lateral displacements of the crust
- raising and lowering of the sea floor
- landslides, liquefaction and tsunami.

The Impact of earthquakes

Impacts are often greater in LEDCs than in MEDCs. In LEDCs building regulations are poor, they are often remote and isolated, emergency planning is non-existent.

Agadir, February 1960 – LEDC
12 000 deaths. 5.7 on the Richter Scale. As a large town it is not surprising that there were a large numbers of deaths. Poorly-built houses collapsed readily.
Problems: In 1962 people were still in tents. But by 1966 75% of Agadir had been rebuilt.
The population of Agadir has since exploded. What happens next time?

Loma Prieta, San Francisco and San Jose, October 1989 – MEDC
62 killed. 7.1 on the Richter Scale.12 000 displaced.
$6 billion of damage. Deaths, injuries and socio-economic disruption were limited because of the state of California's preparedness.
Problems: Marina District wrecked by liquefaction, Cypress Freeway destroyed and Pacific Garden Mall damaged. All the damage was repaired within a few months.

Maharashtra/Khilari, September 1993 – LEDC
25 000 killed. A shallow focus, 6.4 on the Richter Scale, in an area where there had been no tremors for years. 50 villages and towns affected. 7000 homes destroyed. The high death toll relates to the timing of the quake, it happened during the night!
Problems: Dealing with the injured. 90 villages/200 000 people needed new homes. Delay in dealing with the problem. Lack of long-term transfer technologies.

Assisi Earthquake, 1997 – MEDC
10 killed in a densely populated area. 6.0 on the Richter Scale. Many were left homeless.
Many problems had been avoided because 70% had been evacuated, as the earthquake had been predicted. A tented city was established for the homeless and international help and aid was on the scene in hours.
Problems: Damage to the Basilica of St Francis of Assissi and many old frescoes were damaged.

CASE STUDY

Responding to the earthquake hazard

Successful earthquake management depends upon a number of interacting variables.

Physical factors:

- time of day
- depth of focus
- base geology
- duration of the shake
- location of the epicentre.

Human factors:

- building style
- preparedness
- efficiency of the emergency services
- ability to cope and react.

Other factors:

- pinpointing of weaknesses in the infrastructure
- identification of localities prone to liquefaction/folding or faulting
- charting of the recurrence interval.

Loma Prieta Earthquake, October 1989
What could have happened if the quake had been near a population centre?

Actual:	*Hypothetical:*
Epicentre was in the Santa Cruz Mountains, 50 miles away.	Epicentre close to population centre.
Dead 62	Dead up to 4500
Injuries 3757	Injured up to 135 000
Damage $6 billion	Damage $40 billion
Home damage 18 000	One or more airports out of action
Persons displaced 12 000	Bay area difficult to access for days
Bay Bridge was out of action for one month	The city of San Francisco severely damaged

The above case study indicates why planning for an earthquake event is so necessary.

Action can be taken to:

- modify (change) the event – through hazard resistant design, perhaps even controlling the earthquake

- modify a community's vulnerability (reduce losses) – through prediction and warning, through community preparedness and land-use planning

- modify the loss (to the community) (accepting it) – through aid and insurance.

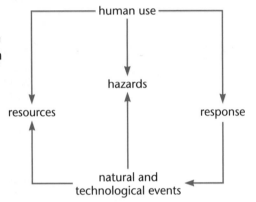

If one looks at the relationship between natural hazards – like earthquakes and volcanoes – there is a clear link between national income levels and deaths. This seems to relate to the size of the hazard, the level of adjustment to the hazard, population density and the perception of the hazard threat.

The need to be well prepared has received a necessary high profile in MEDCs, less so in LEDCs, but evidence suggests (e.g. Philippines) that the situation is improving for the better.

4.2 Rock type and weathering processes

After studying this section you should understand:

- the role that weathering, slope systems and their development have on the landscape
- the effect that different geology's have on surface relief

LEARNING SUMMARY

Rocks and relief

EDEXCEL A   U1

Igneous rocks

Granite is a course-grained intrusive igneous rock consisting of three minerals: quartz (resistant): feldspar (susceptible to chemical weathering) and mica. The largest form of granite structure is batholiths, but it may occur in other forms such as dykes (vertical) and sills (horizontal).

According to Linton's theory of tor formation, granite batholiths formed about 6000 metres below the present-day surface of Dartmoor by the cooling of magma about 450 million years ago. Percolating groundwater caused chemical weathering by hydration and hydrolysis (see next section), especially in zones which are jointed. Subsequent removal of the overlying sedimentary strata has exposed the unweathered remnants of the batholiths as tors, for example Hound Tor. The horizontal divisions between the blocks are pseudo-bedding planes; the result of pressure release as the overburden is reduced by denudation.

The development of tors

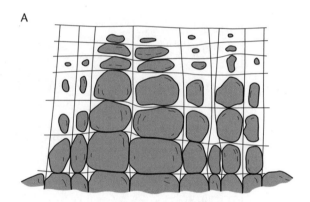

A

B

Sedimentary rocks

Carboniferous limestone such as at Malham in North Yorkshire is composed of calcium carbonate. It displays horizontal bedding and vertical jointing and thus possesses secondary permeability. The best known feature of karst landscapes is the limestone pavement, comprising clints (blocks) and grykes (vertical joints). The chemical weathering process of carbonation enlarges the grykes. Other karstic landforms include sinkholes, caves and resurgent streams. Also typical of limestone areas are dry valleys. The theory attributed to their formation is that during the periglacial periods of the Pleistocene, when the underlying rock was frozen solid, all the water had to run off at the surface, thereby eroding the dry valleys.

A comparison of a chalk (South Downs) and a limestone (Yorkshire Dales) area.

Varying amounts of calcite and dolomite cause the differences between chalk and limestone.

	South Downs	*Yorkshire Dales*
Descriptors	Billowy, rounded and smooth soft profiles.	Broken, scarred and chiselled, a strong angular look.
Structure	Gulled, and frequent coombes. The questa with its escarpments and dip slope.	'Terraced' into beds, pavements, sinkholes and caves.
Water	Chalk is permeable and lacks perennial streams. Flow of rivers is affected by the position of the water table. Many winter-borne streams exist. Streams tend to be widely spaced. Resurgences are rare.	The solubility of limestone means it lacks the dendritic development of the river systems found on chalk. Limestone tends to exhibit few rivers: it is a 'dry' landscape. Resurgences are common.
Slope foot	Sludge and coombe deposits.	Angular shattered debris.

The typical surface features and patterns on chalk and limestone.

CASE STUDY

Chalk – permeable rock at or near surface (The South Downs)

S = Spring

system of back slope branching dry valleys (Dendritic pattern)

Carboniferous limestone – soluble rock at or near surface (The Pennines)

S = Spring

entrenched river valley in impermeable rocks below upland of soluble rock intermittent drainage (?)

Metamorphic rocks

Metamorphic rocks are formed from either igneous or sedimentary rocks by renewed heat and/or pressure.

Examples include marble from limestone and slate from shale.
All metamorphic rocks are impervious (they will not allow the infiltration of water).

Weathering: processes of disintegration

AQA A	A2	EDEXCEL A	U1
AQA B	UA	WJEC	A2
OCR A	U1	NICCEA	U1

Igneous rocks form in the Earth's interior, where pressures and temperatures are very different from those at the Earth's surface. Forty kilometres below the surface pressures may be as high as 12 tonnes/cm^2 and temperatures may be as high as 600°C. The chemical and physical characteristics of rocks reflect these conditions. At the surface, where atmospheric pressure is only about 6.7 kg/cm^2 and temperatures rarely exceed 35°C, most rocks are physically and chemically unstable; they break up by being weathered.

Rock weathering is the disintegration and decomposition of rocks, *in situ* (in one place), by natural agents at or near the Earth's surface. It is different from erosion, which requires moving agents.

Denudation is the general term given to the wearing away of the Earth's surface by weathering *and* erosion. There are three main types of weathering.

Physical weathering

Frost shattering (freeze-thaw) relies on water in cracks subjected to many freeze-thaw cycles (fluctuations in temperature around 0°C). Water expands by 9.6% of its volume as it freezes, exerting a stress on the rock of up to 2100 kg/cm^2 at –22°C. Though frost shattering has been observed happening at 14 kg/cm^2, strong rocks, if jointed, will split at 100 kg/cm^2. Frost action is at a maximum in rocks that do not drain freely. Chalk, with its pore spaces and moisture retaining properties, disintegrates readily in freezing conditions. In Switzerland the process causes an annual loss of 1 to 2 mm/yr. Mount Kenya in Africa has been dramatically affected by frost shattering! Periglacial areas experience very little frost shattering.

Salt crystallisation is caused by the crystallisation of supersaturated solutions of salts (e.g. sodium chloride) occupying fissures and pore spaces within rocks. As the crystals grow, pressure causes surface scaling or granular disintegration. It also produces cavernous weathering (honeycombing of the surface) and contributes to the production of weathering pits and **tafoni**.

Insolation weathering (exfoliation) occurs in desert environments with a large diurnal range in temperature (the difference between maximum daytime and minimum night-time temperatures). Because rock is a poor conductor of heat, the outer layers are subject to alternate expansion (due to intense daytime heating) and contraction (due to rapid cooling at night), while the inner layers remain cold. The outer layers eventually peel off. Water, in tiny amounts, plays an important part in the process. The effect of **exfoliation** is insignificant (0.5 mm/10 000 yr).

Pressure release (**dilatation**) occurs in many rocks, especially intrusive jointed granites which have developed at considerable pressure and depth. The confining pressure increases the strength of the rocks. If these rocks are exposed to the atmosphere at a later date (due to removal of the overburden by, for example, glacial erosion) then there will be a substantial release of pressure (at right angles) which weakens the rock, allowing other agents to enter it and other processes to occur. Where cracks develop parallel to the surface, the process of sheeting causes the outer layers of the rock to peel away (along what are called pseudo-bedding planes). Pressure release is thought to be the dominant process in the formation of **inselbergs**, and certainly perpetuates them. (Inselbergs are bare, rounded, steep-sided and dome-shaped features.)

Chemical weathering

This contributes to the disintegration of rocks by:

- weakening the coherence between minerals
- attacking the cements, e.g. in sandstone

- forming solutions, which are washed out by rain, making the rock porous and so ready to crumble by granular disintegration
- causing the formation of alteration products.

Hydration occurs in all rocks. Certain crystals grow in size due to the addition of water creating stresses in the rock, which may eventually crumble and disintegrate (granular disintegration). **Hydrolysis** occurs in rocks containing the mineral feldspar (e.g. granite). Hydrolysis is a chemical reaction between water and the hydrogen ions in the rock. The feldspar in granite decomposes to form kaolin which is washed away. The laterites of the tropics, when rich in aluminium, give up bauxite.

Oxidation occurs in rocks containing iron compounds. Oxygen dissolved in water reacts with the iron to form oxides and hydroxides. This manifests itself in a brownish or yellowish staining of the rock surface which ultimately disintegrates. **Carbonation** occurs in rocks containing calcium carbonate. Carbon dioxide in the atmosphere is absorbed by rain, creating a weak carbonic acid. This chemically converts the insoluble calcium carbonate (found in limestone and chalk) into soluble calcium bicarbonate, which is dissolved and carried away in solution.

Biological weathering

Plant roots widen cracks in rock faces; rotting plants create humic acid, which chemically decomposes the rock; animals produce ureic acid, which chemically decomposes the rock. In the UK there are 150 000 worms/acre, they move 10 to 15 tonnes of soil and other weathered rock fragments per year.

Slopes and slope processes

AQA A	A2	EDEXCEL A	U1
OCR A	U1	EDEXCEL B	A2
OCR B	A2		

Slopes form much of the Earth's surface and influence most aspects of geography including land use, farming, transport networks, rivers and their basins. Many slopes are unstable and changing all of the time, sometimes rapidly, while the shallowest slopes 'creep'. Slope shapes are mostly due to past slope movement.

Weathering processes produce debris, which is transported downslope by processes of mass movement. Mass movement processes may be sub-divided into rapid and slow, wet and dry. The type of process involved and its velocity depend on many factors including the angle of slope, the amount of water present in the waste material, the degree and type of vegetation, and human activity.

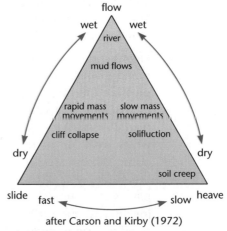

after Carson and Kirby (1972)

Types of movement

Soil creep
This is very slow, does not require a lot of water and is the result of surface material being 'heaved' down slope.

Solifluction

Occurs in periglacial climates. The ground tends to be wet (in the summer) and movements are relatively fast.

Cliff collapse

This rapid slope movement occurs on steep slopes.

Mudflows

When very wet this occurs very rapidly.

Slope movements on soft rocks occur rapidly, and differ from the movements that occur on hard rock. The most common mass movements are those that occur through gravity on soft rocks. Most of these movements are rapid and occur over a clearly defined boundary, called a slip surface or shear plane.

Types of rapid movement include the following.

Translational slides

The slip surface is parallel to the slope and they are not very deep. Heavy rainfall over a period of time can start them off.

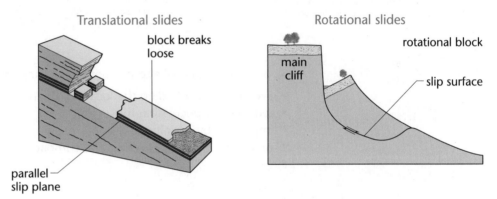

Translational slides — block breaks loose — parallel slip plane

Rotational slides — main cliff — rotational block — slip surface

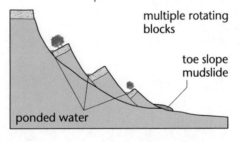

Multiple rotational slides — multiple rotating blocks — toe slope mudslide — ponded water

Rotational slides

Large blocks of material rotate over a curved slip surface. The ground surface is left tilting back towards the main cliff. Often a number of blocks rotate down the hill giving a step-like slope profile.

Mudslides

These usually occur in debris that has already slipped downslope. High water content means that they can flow at fast speeds for long periods.

The loose debris at the foot of a slope is termed **talus**. The form (profile/angle) of a slope depends on the balance between subaerial supply of talus by weathering and mass movement processes and its removal at the cliff foot by, for example, a river or the sea. Where subaerial supply exceeds basal removal, a low-angled, degraded profile develops. Where basal removal exceeds subaerial supply, a more vertical profile is maintained.

Slope development theories

	slope decline (W.M. Davis, 1899)	slope replacement (W. Penck, 1924)	parallel retreat (L.C. King, 1948, 1957)
climate	humid climates	tropical areas	semi-arid landscapes
description of slope	Steepest slopes at beginning of process with a progressively decreasing angle in time to give a convex upper slope and a concave lower slope.	The maximum angle decreases as the gentler lower slopes erode back to replace the steeper ones, giving a concave central portion to that slope.	The maximum angle remains constant as do all slope facets, apart from the lower one which increases in concavity.
how the slope develops	(a) slope decline stage 3 stage 2 stage 1 stage 4 convex curve concave curve watershed worn down by stage 4 land has been worn down into a convex-concave slope	(b) slope replacement stage 3 stage 2 stage 1 A A A C B C B C B B = talus-scree slope and will replace slope A C will eventually replace slope B	(c) slope retreat stage stage stage stage 4 3 2 1 convex free face concave debris slope pediment (can be removed by flash floods)

Slope movements have to be controlled. They can be:

- avoided, by controlling the location, timing and nature of development
- reduced, by decreasing the angle
- improved, by draining and using retaining structures
- protected, through covering and compaction.

1

Two types of boundary between tectonic plates are illustrated below

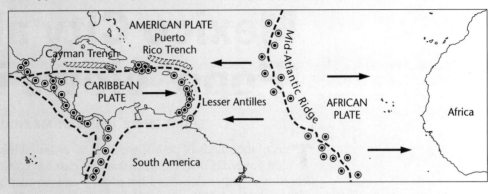

← Direction of plate movement   ⟋⟋⟋⟋⟋ Ocean trenches

------ Plate boundary   ◉   Earthquake epicentres

(a) (i) Name the type of boundary found adjacent to the Lesser Antilles.   [2]

Destructive/consuming or equivalent.

(ii) Draw and label a simple section to show what is happening along this boundary.   [4]

Needs a simple diagram like the one in the tectonic chapter.

(b) (i) Name the type of boundary found along the mid-Atlantic ridge.   [1]

A constructive/accreting boundary or equivalent.

(ii) Give **two** reasons why the American plate slides beneath the Caribbean.   [2]

* *Differences in density of plates.*
* *Atlantic plate runs under the near-static Caribbean plate*
* *comparative size of the plates.*

(c) (i) Describe the processes which are causing the African and American plates to move away from each other.   [4]

* *Convection currents and magmatic plumes*
* *subduction drag and slab pull.*

(ii) Give three pieces of evidence for this movement.   [3]

* *The shape of the continents suggest they were once joined.*
* *studies of rock that make up the upper asthenosphere*
* *radioactive decay.*
* *Rock types and geology—in NE Brazil/Liberia and Guinea, South Africa/Argentina.*

(d) Why are so many earthquake epicentres beneath the West Indian islands.   [2]

Great stresses at the subducted limb of a destructive plate boundary. Reabsorption and slab jerk cause earthquakes in the Benioff zone. The West Indian Islands lie directly above the subducted limb of the Lesser Antilles.

(e) Explain the origin of the islands of the Lesser Antilles.   [4]

The explanation should focus on Island Arc Theory/processes.

(Edexcel)

Practice examination questions

Some boards are using article-resourced questions – try one!

1

Mexico City alert as 'Popo' blows its top

FROM CHRIS ASPIN IN MEXICO CITY

Tens of thousands of people were on alert yesterday after a huge volcano near Mexico City belched red-hot rocks and clouds of ash in its biggest eruption in 72 years.

Ash and acrid gases showered down on much of Mexico City late on Monday from the snow-capped, 17,890ft Popocatepetl volcano, forcing a temporary closure of the international airport.

Later, officials said activity in the volcano appeared to be subsiding and one runway of the airport was reopened, but yesterday morning television soon showed Popocatepetl spewing another mushroom cloud. "The [volcano tracking] system will remain in the same state of alert as yesterday," Manuel Bartlett, Governor of Puebla State, said.

Troops were rushed to 30 villages at the foot of the volcano in preparation for an evacuation.

Winds blew the fine grey ash as far east as the Caribbean port of Veracruz and residents in Mexico City complained of burning eyes and throats. "It feels like sand. It gets into your throat and your teeth. We've never seen anything like this," Maria del Carmen Ariaga said as she watched the blanket of ash fall.

One official said at least 193,000 people were ear-marked for evacuation from Puebla city alone.

Experts said "Popo", as the mountain is affectionately known, had registered its most violent activity since 1925. The volcano, whose name means "smoking mountain" in the Nahuatl Indian language has blown ash and steam on many occasions this year. But Monday's explosion was the first time that the grit-like ash had rained down on Mexico City.

(Reuters)

(a) Volcanoes like Popocatepetl rarely release magma onto the surface but they do produce a range of potential pyroclastics that can kill.
Describe the source/cause and effects of at least two of these pyroclastic events. The article may aid your answer. [8]

(b) Explain why the slopes of active volcanoes like Popocatepetl continue to attract settlement. [8]

(c) For one other natural hazard you have studied suggest how people adjust, modify and cope with the hazard. [9]

Practice examination questions (continued)

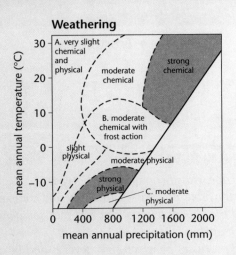

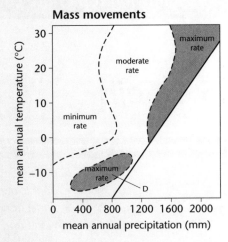

2 (a) What is meant by the term 'weathering'? [2]

(b) Name and explain one possible type of physical weathering process found at location A. [2]

(c) Explain why frost weathering is likely to be more intense at location B than at A. [2]

(d) What is meant by the term 'mass movement'? [2]

(e) Name one possible type of mass movement at location D. [1]

(f) Explain how this type of mass movement is influenced by the climate at location D. [2]

(g) Apart from climate, state and explain two other factors which may influence rates of mass movement. [4]

[OCR]

Chapter 5

Soils and ecosystems

The following topics are covered in this chapter:

- *Soils*
- *Ecosystems*

✓ 5.1 Soils

After studying this section you should be able to:

- *define, describe, classify and locate soil types*
- *appreciate that soils are highly organised and complex ecosystems*
- *understand the vital and important role that soil plays in the life of the world and mankind*
- *understand processes by which soils develop and be aware of how to manage them sustainably*

Formation of soil

AQA A	U1	EDEXCEL A	A2
AQA B	A2	EDEXCEL B	A2
OCR A	U1	WJEC	U1
		NICCEA	U1

> The study of soil is called Pedology.

> 1cm of soil takes between 100 to 1000 years to develop.
> H. Jenny represented these factors in his formula
> $s = f(cl,o,r,p,t)$
> (s = soil, f = factors, cl = climate, o = organic inputs, r = relief, p = parent material, t = time).

Soil is a vital, but often forgotten, resource.

It is a residual layer of weathered material that has accumulated over a long period, sometimes known as the **regolith**. Geographers regard soil as an ecosystem; it has all the major functioning components of a dynamic ecosystem. The principle components of the soil system are water, mineral matter, air, organic material and biota; the proportions of each component are shown to the right. These components form part of a complex inter-relationship with a number of other factors, including parent material, climate, relief and time.

The soil profile

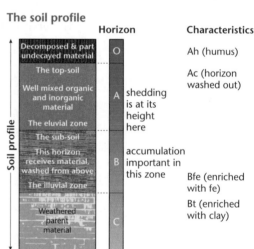

> Water is lost from the soil system from the top soil and from sub-surface layers by evapotranspiration and via run-off.

Transformation processes

Water changes both the chemical composition and form of soil materials. Bacteria and insects break down and transforms litter into humus and is of major importance in transforming the mineral fraction through weathering. The processes of: **solution**, in the CO_2 rich atmosphere of the soil; **hydration**, whereby minerals react in the presence of water by destroying crystal structures; **hydrolysis**, the breakdown of minerals by hydrogen and hydroxyl ions derived from water, and **oxidation** and **reduction** all contribute to soil formation.

The role of the principle soil components

> The role of water is without question!

Water

Water is stored temporarily in soil pores. Water is important in terms of vertical (upwards and downwards) and lateral movements (on slopes) in the soil. These movements are known as **transfers**. Chemical changes, brought about by the

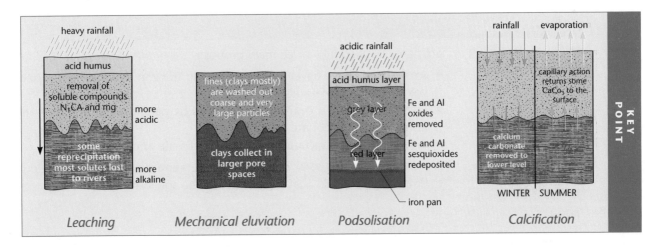

Leaching | Mechanical eluviation | Podsolisation | Calcification

presence of water, are known as **transformations**. Together these are the principal **pedogenetic processes**.

Organic matter

This is the remains of soil fauna and flora decomposed to form humus. Humus is found in the upper parts of the A horizon. There are often three distinct layers or stages of decomposition in the 'humus', i.e. litter layer, fermentation layer and humification layer. Different soils have different humus coverings, i.e. **mor** on acid soils and **mull** on alkaline soils. Organic material contributes nutrients to the soil. Clay and humus particles unite to form negatively charged molecules which attract positively charged plant food in the soil. This relationship forms the so called clay-humus molecule.

Mineral matter

This makes up the largest part of most soils and contributes the abiotic substances to the system. Mineral matter influences the size of soil particles. Three mineral fractions are identified based on the diameter sizes of particles, sand (2–0.05 mm), silt (0.05–0.002 mm) and clay (less than 0.002 mm). The proportions of sand, silt and clay determine soil texture (see soil texture triangle). Of the three fractions clay is probably the most important because it is chemically active. All clay particles carry a negative charge and so repel one another, they are said to be **colloidal**. Clay particles are said to **flocculate** when they finally come together. This process is often accelerated by adding lime or manure to improve aeration and drainage.

The arrangement of mineral matter determines soil structure. Soil particles cling together, or aggregate, to form **peds** (five shapes are recognised). Peds not only contain plant nutrients but also allow for aeration and drainage.

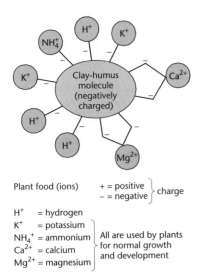

Plant food (ions)   + = positive }
                      – = negative } charge

H⁺ = hydrogen
K⁺ = potassium
NH₄⁺ = ammonium   } All are used by plants
Ca²⁺ = calcium   } for normal growth
Mg²⁺ = magnesium   } and development

> The ped structure determines the soil fabric or percentage of pore spaces available.

Appearance of aggregates (peds)					
Type of structure	Crumb	Platy	Blocky	Prismatic	Columnar

Soil classification

AQA A	U1	EDEXCEL A	A2
AQA B	A2	EDEXCEL B	A2
OCR A	U1	WJEC	U1
		NICCEA	U1

Zonal soils are the ones to learn at AS. You should expect questions on podsols, chernozems, brown earths and tropical soils. Possibly gleys and salinised soils too.

Soils are classified into three main groups:

- **Azonal soils** have developed recently and horizon development is poor, e.g. volcanic and alluvial soils.
- **Intra zonal soils**, are strongly influenced by parent material, e.g. rendzinas from calcium carbonate.
- **Zonal soils** are associated with specific climate and vegetation.

A comparison of the soils, climate and vegetation on the diagrams below serves to emphasise the links that exist between the soil formation factors.

Podsol

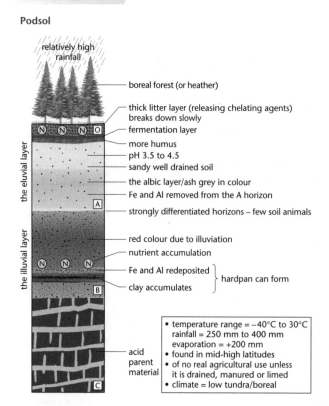

relatively high rainfall

- boreal forest (or heather)
- thick litter layer (releasing chelating agents) breaks down slowly
- fermentation layer
- more humus
- pH 3.5 to 4.5
- sandy well drained soil
- the albic layer/ash grey in colour
- Fe and Al removed from the A horizon
- strongly differentiated horizons – few soil animals
- red colour due to illuviation
- nutrient accumulation
- Fe and Al redeposited } hardpan can form
- clay accumulates
- acid parent material

the eluvial layer / the illuvial layer

- temperature range = −40°C to 30°C
  rainfall = 250 mm to 400 mm
  evaporation = +200 mm
- found in mid-high latitudes
- of no real agricultural use unless it is drained, manured or limed
- climate = low tundra/boreal

Brown earth

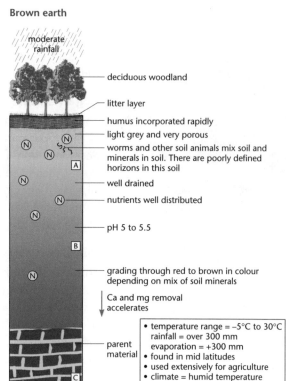

moderate rainfall

- deciduous woodland
- litter layer
- humus incorporated rapidly
- light grey and very porous
- worms and other soil animals mix soil and minerals in soil. There are poorly defined horizons in this soil
- well drained
- nutrients well distributed
- pH 5 to 5.5
- grading through red to brown in colour depending on mix of soil minerals
- Ca and mg removal accelerates
- parent material

- temperature range = −5°C to 30°C
  rainfall = over 300 mm
  evaporation = +300 mm
- found in mid latitudes
- used extensively for agriculture
- climate = humid temperature

Chernozem

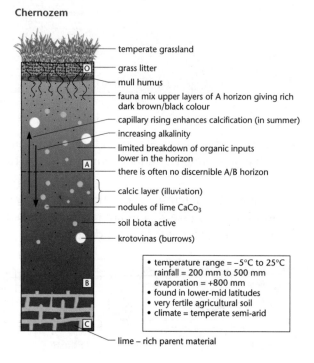

- temperate grassland
- grass litter
- mull humus
- fauna mix upper layers of A horizon giving rich dark brown/black colour
- capillary rising enhances calcification (in summer)
- increasing alkalinity
- limited breakdown of organic inputs lower in the horizon
- there is often no discernible A/B horizon
- calcic layer (illuviation)
- nodules of lime CaCo₃
- soil biota active
- krotovinas (burrows)
- lime – rich parent material

- temperature range = −5°C to 25°C
  rainfall = 200 mm to 500 mm
  evaporation = +800 mm
- found in lower-mid latitudes
- very fertile agricultural soil
- climate = temperate semi-arid

Tropical soil

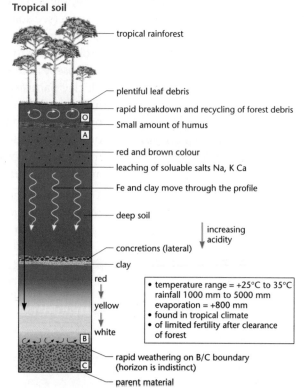

- tropical rainforest
- plentiful leaf debris
- rapid breakdown and recycling of forest debris
- Small amount of humus
- red and brown colour
- leaching of soluable salts Na, K Ca
- Fe and clay move through the profile
- deep soil
- increasing acidity
- concretions (lateral)
- clay
- red → yellow → white
- rapid weathering on B/C boundary (horizon is indistinct)
- parent material

- temperature range = +25°C to 35°C
  rainfall 1000 mm to 5000 mm
  evaporation = +800 mm
- found in tropical climate
- of limited fertility after clearance of forest

Reasons for variation down the length of the hillslope include:
- drainage condition
- transport of eroded materials
- transporting by leaching.

Catenas

Soil development is often related to the shape, lie and aspect of the land. Regular sequences of soil occur on slopes. These predictable patterns are called catenas. Typically, podsols form on the well-drained upper slopes, brown earths in the intermediate central area and gleys on the water-logged slope foot.

Soil erosion and management

AQA A	U1	EDEXCEL A	A2
AQA B	A2	EDEXCEL B	A2
OCR A	U1	WJEC	U1
		NICCEA	U1

About 12% of the total land area of the Earth's 13 billion hectares is cultivated!

Soil erosion

Misuse of soil on the whole reflects failure to understand and manage it. The result of this misuse is soil erosion, the accelerated process of natural landscape evolution! Soil erosion directly relates to overgrazing, intensive cultivation and deforestation, but this is only part of the problem. Soil degradation is much more serious: this involves misuse but also fertility decline.

Management of soils

In Norfolk, UK, soil erosion is related to overfine soil texture, increased surface runoff, overlarge fields with no hedges and intensive farming. 44 000 tonnes of soil is lost per year.

Soil management involves introducing techniques which maintain productivity without causing environmental damage. The best cure for accelerated soil erosion is to mimic the usual or natural situation either by:

- intercropping
- leaving crop residues in place after harvesting
- improving and maintaining soil structure limits removal by water and wind

or by:

- introducing physical measures, such as windbreaks, shelter belts or terracing
- using appropriate crop practices
- planning careful future use of fertilisers
- limiting, where feasible, deforestation
- implementing appropriate agriculture and agricultural techniques.

Iceland's degraded soils

The extent of Iceland's soil degradation is perhaps surprising given its humid climate. Most of the world's most severely degraded areas are found in the arid regions of the world. Like other degraded areas, Iceland displays all those features that lead to accelerated soil erosion. Soon after settlement, rapid population growth led to overuse of what is a very fragile ecosystem. Grazing, cutting and burning destroyed vegetation cover and allowed the processes of mass movement, combined with ferocious surface winds and high rainfall, to exacerbate the soil erosion problem. Nearly 60 000 km^2 of Iceland is now classified as desert, or is of little agricultural use. Cooling of the climate along with Iceland's volatile landscape means its soils will never recover.

CASE STUDY

5.2 Ecosystems

After studying this section you should understand:

- that living things perform basic functions: feeding, growing, respiring, etc.
- that plants build tissue through photosynthesis
- that vegetational biomes display certain distributions, related to climate
- that vegetational systems display regular patters of succession
- the functions of a small scale ecosystem and one major biome

Structures within ecosystems

AQA A	U1	EDEXCEL A	A2
AQA B	UB	EDEXCEL B	A2
OCR A	U1	WJEC	U1
OCR B	A2	NICCEA	U1

GPP is the total amount of energy fixed by green plants.

NPP contributes to the production of new plant tissue.

NPP = GPP – Respiration.

An ecosystem is composed of living and non-living components interacting to produce a stable system.

Cycling of energy

This is perhaps the most obvious feature or process in the interactive ecosystem. Energy cannot be created or destroyed but can be transferred. Primary production via photosynthesis fixes energy from the sun into plants. The rate of energy fixing is expressed in kg/m²/year (or equivalent), and is fixed either as **GPP** (gross primary production) or as **NPP** (net primary production).

Energy is lost/released through respiration in the photosynthetic process. Energy is moved or transferred in **food chains**. Whenever energy is exchanged in a food chain, a 'stage' or **trophic level** is reached. There are few food chains with more than five trophic levels because of this energy loss or exchange. The numbers of individuals, biomass and productivity decreases as trophic levels four and five are reached.

Primary producers = autotrophs.

These diagrams often form the basis of AS questions.

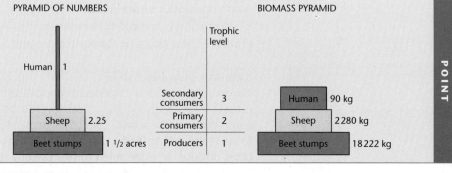

K E Y
P O I N T

Simple food chains seldom exist, it is more realistic to display energy flows in a food **web**.

Cycling of nutrients

This is the second of the processes in an ecosystem. Nutrient cycling moves chemical elements out of the environment to organisms and then back again to the environment. Nutrients are stored in compartments and are recycled via pathways, as shown in the diagram.

Stores are not in a static condition. As nutrient amounts increase or decrease, store sizes vary. N.B. they can be drawn any shape!

Humans can affect the cycling of nutrients in ecosystems by misusing or overusing the land.

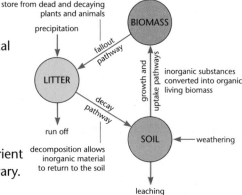

Biomes: the distribution of vegetation on a global scale

AQA A	U1	EDEXCEL A	A2
AQA B	UB	EDEXCEL B	A2
OCR A	U1	WJEC	U1
		NICCEA	U1

Soil factors are often referred to as 'edaphic factors'.

There is a very close link/relationship between climates, soil and vegetation. This zonal relationship can be explained using the concept of the biome (named after the vegetation found within a zone). Four factors determine biomal distribution: soil, relief, biota and of course climate.

Relief is to do with the shape of the land surface.

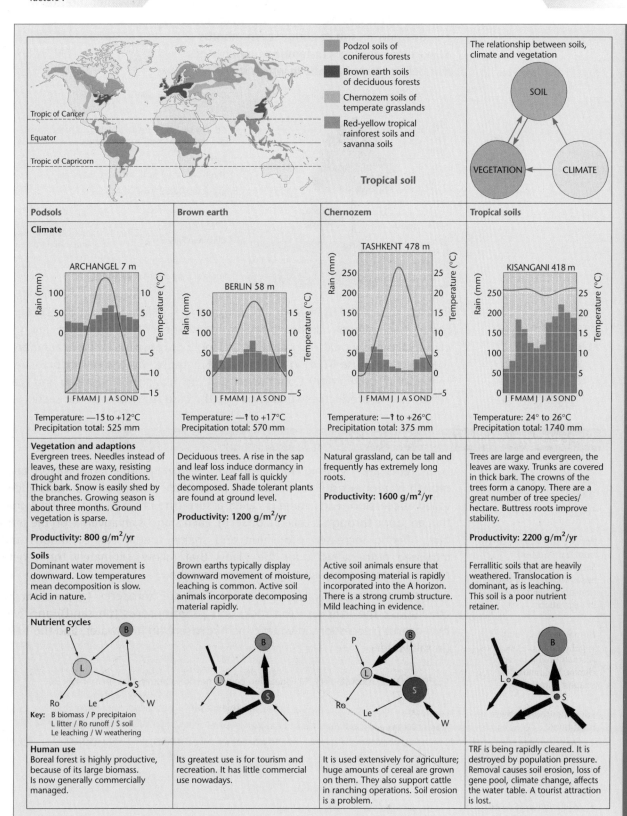

Podzol soils of coniferous forests

Brown earth soils of deciduous forests

Chernozem soils of temperate grasslands

Red-yellow tropical rainforest soils and savanna soils

The relationship between soils, climate and vegetation

Tropical soil

Podsols	Brown earth	Chernozem	Tropical soils
Climate			
ARCHANGEL 7 m	BERLIN 58 m	TASHKENT 478 m	KISANGANI 418 m
Temperature: —15 to +12°C Precipitation total: 525 mm	Temperature: —1 to +17°C Precipitation total: 570 mm	Temperature: —1 to +26°C Precipitation total: 375 mm	Temperature: 24° to 26°C Precipitation total: 1740 mm
Vegetation and adaptions Evergreen trees. Needles instead of leaves, these are waxy, resisting drought and frozen conditions. Thick bark. Snow is easily shed by the branches. Growing season is about three months. Ground vegetation is sparse. **Productivity: 800 g/m²/yr**	Deciduous trees. A rise in the sap and leaf loss induce dormancy in the winter. Leaf fall is quickly decomposed. Shade tolerant plants are found at ground level. **Productivity: 1200 g/m²/yr**	Natural grassland, can be tall and frequently has extremely long roots. **Productivity: 1600 g/m²/yr**	Trees are large and evergreen, the leaves are waxy. Trunks are covered in thick bark. The crowns of the trees form a canopy. There are a great number of tree species/ hectare. Buttress roots improve stability. **Productivity: 2200 g/m²/yr**
Soils Dominant water movement is downward. Low temperatures mean decomposition is slow. Acid in nature.	Brown earths typically display downward movement of moisture, leaching is common. Active soil animals incorporate decomposing material rapidly.	Active soil animals ensure that decomposing material is rapidly incorporated into the A horizon. There is a strong crumb structure. Mild leaching in evidence.	Ferrallitic soils that are heavily weathered. Translocation is dominant, as is leaching. This soil is a poor nutrient retainer.
Nutrient cycles			
Human use Boreal forest is highly productive, because of its large biomass. Is now generally commercially managed.	Its greatest use is for tourism and recreation. It has little commercial use nowadays.	It is used extensively for agriculture; huge amounts of cereal are grown on them. They also support cattle in ranching operations. Soil erosion is a problem.	TRF is being rapidly cleared. It is destroyed by population pressure. Removal causes soil erosion, loss of gene pool, climate change, affects the water table. A tourist attraction is lost.

Key: B biomass / P precipitaion L litter / Ro runoff / S soil Le leaching / W weathering

KEY POINT

Small-scale ecosystems in the UK

The broadleaf temperate deciduous forest (BTDF)

Present plant species in the BTDF date on the whole from the last ice-age. Since its first appearance the BTDF has been markedly depleted, especially near the major cities in the UK. In the world as a whole, 19% of the biomass by biome is BTDF. BTDF displays characteristic features:

- leaf loss, a form of dormancy in response to the cold winter months when there is little soil water and low sunlight/energy
- a growing season of about seven months
- extreme sensitivity to temperature changes
- a definite tree line, above which it will not grow.

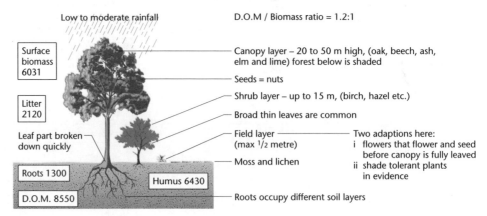

Principle soils of the BTDF are the brown earths. Significant uses of the BTDF include game management, timber production, as shelter-belts and recreation. Most conservationists would advocate that this multi-purpose use is, if managed properly, the best way to use BTDF. In 1992, Forest Enterprise was one of the 'branches' that came out of the Forestry Commission split. Its mission is to develop and maintain attractive and productive woodlands (see the case study later in this section).

Succession and climax: the changing ecosystem

AQA A	U1	EDEXCEL A	A2
AQA B	UB	EDEXCEL B	A2
OCR A	U1	WJEC	U1
OCR B	A2	NICCEA	U1

Primary succession starts from bare ground. Secondary succession arises on land that had previously been covered in vegetation.

Human activity holds vegetation in plagioclimax. Animals hold it in biotic climax. Soil holds it in edaphic climax.

For dunes to form there must be :

- a replenishable source of sand
- regular onshore winds
- obstacles on the beach to catch the sand
- vegetation to stabilise the dunes.

Ecosystems are dynamic and continually evolving. After colonisation and the growth of **pioneer** species begins the predictable pattern of vegetational change called **succession**. Each stage is called a **sere** (the chain of vegetational changes that are gone through is called a **prisere**). Each stage is an advance on the previous one, offering increased protection and shelter, better soil conditions and an increased nutrient stock for the plants that follow. Ultimately the vegetation reaches its most complex stage, **climax** is reached.

Four types of succession/prisere exists. Two are **xeroseres** (successions in dry conditions): the **lithosere** (on rock) and **psammoseres** (on sand), and two are **hydroseres** (successions in water): the **hydrosere** (in freshwater) and the **halosere** (in salt water).

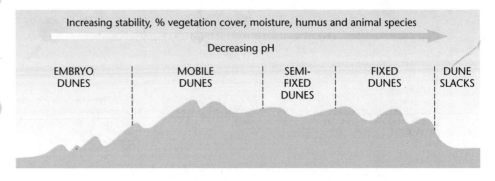

The development of a psammosere in West Wales

Ynyslas Dunes in West Wales are a typical example of a psammosere.

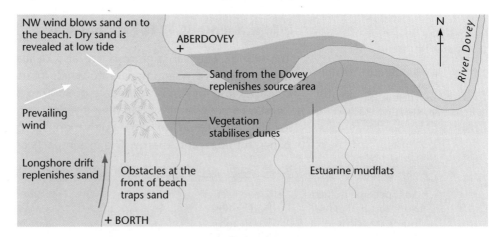

Embryo dunes

An extremely difficult environment. Sand is dry, lacking in nutrients, has poor water retention, high alkaline pH and high salt content. Pioneer plants have to be **xerophytic** (drought tolerant) and **halophytic** (salt tolerant). Typical plants include sea rocket and sea couch. Few animals colonise the embryo dunes.

Mobile dunes

Areas of exposed sand reshaped by the wind. **Marram grass** is the dominant vegetational type. It is tolerant of the extremely dry conditions and begins to stabilise dune systems. Groundsel and sea spurge are also common. There is still no real colonisation by animals.

Semi-fixed dunes

These have better conditions for plants. Soils begin to develop. Mosses and lichens start to colonise the area. A range of flowering plants appear: increased nitrogen fixation aids plant development. Animals are common, from rabbits through to invertebrates.

Fixed dunes

These are the most stable of the dunes and are covered in vegetation. Plant growth is at its maximum as conditions are ideal. Humus incorporation into the soil begins.

Dune slacks

These are hollows formed by erosion by wind until the water table is revealed. These can be rich in different varieties of vegetation.

Pioneer plants use a variety of techniques to conserve water:

1 waxy leaves
2 external roots
3 physiological drought.

Marram has rolled leaves, few stoma and deep tap roots to collect water.

Remember: primary succession goes through the following stages:

1 colonisation
2 establishment
3 competition
4 stabilisation
5 climax.

Management of ecosystems

AQA A	U1	EDEXCEL A	A2
AQA B	UB	EDEXCEL B	A2
OCR A	U1	WJEC	U1
OCR B	A2	NICCEA	U1

Management at the local scale

Ecosystems need to be managed at a local scale. Dunes perform an important role in protecting the coastline but their soils and vegetation are fragile: trampling by humans, the burrowing and grazing by rabbits and other animals destroy the thin mantle of regolith revealing sand. This can lead to blowouts, massive deflation hollows on dune crests. Management techniques include:

- moveable board walks, zig-zagged to deter motorbikes and other vehicles
- low fencing to deter 'wanderers' and to funnel visitors
- advice boards and direction arrows to help visitors make the most of their tour
- wardens, tolls and admission fees to control numbers of visitors

- designated car parking
- 'sacrificing' small areas to visitors in some dune systems
- controlling animals by culling, poisoning and shooting.

Management at the global scale

> It will all be gone in less than 100 years at this rate!! Such destruction only generates 2% of Brazil's total GDP! (in 1998).

The world's tropical rainforests are also delicate and important to people. TRF deteriorates at a phenomenal rate when interfered with. At present the top ten countries lose rainforest at the rate of 3000 to 35 000 km²/yr; that is 1.5% of total forest area per year. It is cleared for a variety of reasons: to allow access to minerals, for HEP production, to allow cultivation, for cattle ranching, and for timber for paper, furniture and energy. The five impacts of deforestation are:

Plants and animals	*The water cycle*	*Landscape*	*Climate*	*Population*
• Soils become nutrient impoverished as the nutrient cycle is breached. • Species are lost and the genetic stock/potential is lost. • Niches are lost. • Fertile soil is lost. • Seeds don't germinate.	• Greater sediment load causes silting and flooding. • Runoff increases with no vegetation cover. • Quality of the water is affected.	• Soil is lost. • Gullying and sheet wash is more common. • Duricrust forms an infertile hardpan.	• Carbon release to the atmosphere accelerates global warming. • Rainfall starts to decrease. • Humidity drops. • Changes occur in daily temperatures. • Albedo effects on bare soil.	• Indigenous people lose their homes and way of life. • Exposure to outside influences allows disease to spread.

> Sustainability occurs when exploitation is not greater than the ability of a system to replace itself.

TRF can be managed sustainably to meet the needs of plants, animals and indigenous populations and to contribute to the economic development of countries.

KEY POINT

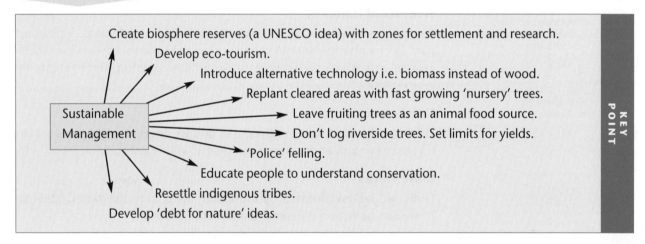

Sustainable Management
- Create biosphere reserves (a UNESCO idea) with zones for settlement and research.
- Develop eco-tourism.
- Introduce alternative technology i.e. biomass instead of wood.
- Replant cleared areas with fast growing 'nursery' trees.
- Leave fruiting trees as an animal food source.
- Don't log riverside trees. Set limits for yields.
- 'Police' felling.
- Educate people to understand conservation.
- Resettle indigenous tribes.
- Develop 'debt for nature' ideas.

The Community Forest Organisation in the UK

Since 1991, 3500 hectares of trees have been planted in twelve community forests in the UK, reviving areas of damaged and derelict land, mostly on the edges of cities. All CFOs have prepared detailed plans that will take them into the middle of the next century. As an example, the Greenwood Community Forest (incorporating Sherwood Forest) in Nottinghamshire, provides opportunities for commercial forestry, access, community involvement and education.

CASE STUDY

Samples question and model answer

1

The graph below shows the average values for the composition of soil.

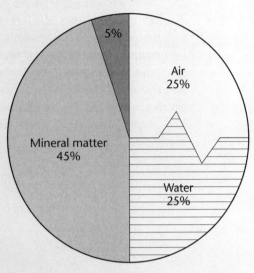

Soil composition

Adapted from Bridges, *World Soils*, CUP 1979

The first few questions offer a way into the question as a whole; ensure you respond to such questions positively.

(a) Complete the answers in the spaces provided: [3]

   (i) Name the constituent which makes up an average of 5% of the soil composition.

      Humus

   (ii) What is the main source of the mineral matter in the soil?

      Parent material

   (iii) Name one type of soil which has a significantly higher than average percentage of water.

      Peat/Gley

(b) Suggest why the line used in the graph to separate water and air was not drawn as a straight line. [2]

Not a bad answer though it did overrun. Try to stick to the line allowance given/offered. Think about what you want to say before committing pen to paper.

    The water and air content is not always the same, it is often of different proportions due to climatic conditions, e.g. in a drought there will be a higher air content than water content and vice versa for flood conditions. The overall water/air content is affected by the pore size.

(c) Outline the reasons why soils in the humid tropics are usually:

   (i) red or yellow in colour [2]

Leaching missed.

    The colour is from sesquioxides. Fe and Al oxides are leached out.

   (ii) more than twice as deep as soils in temperate latitudes. [2]

    The high temperatures and moisture levels create high levels of chemical weathering causing the topical soils to form very quickly.

Sample question and model answer (continued)

Focus is for the most part on drainage. There is little development of either the water/air or clay-humus-complex idea.

Plenty of description, more reasons needed. These questions need answers which are well organised. Theory has to be integrated into your account. Balance is necessary, the question is in two parts. Work to the command words. It pays to learn definitions!

(d) State what is meant by soil texture and explain why it is so important in determining the agricultural potential of a soil. [6]

Soil texture is related to the proportion of sand, silt and clay in a soil. The proportions produce either a coarse or fine textured soil. The texture produces different pore sizes and particle sizes so the aeration and moisture content is affected. The importance of texture to farming, again depends totally on pore size. For example, sandy soils are well drained and the looseness of the particles allows root penetration. Clay soils hold a lot of water but are difficult to penetrate due to their compactness. Silty soils often lack mineral and organic matter as there is little humus (which holds the soil together).

AQA

Practice examination questions

1 The diagram below shows a pedologists ideal profile!

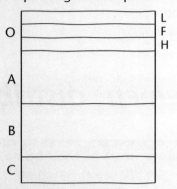

(a) Identify the horizons: A and B. [1]

Identify the layers: F and H. [1]

(b) How might the structure of the soil affect two physical properties of the soil? [2]

Why does the thickness of the O horizon vary through the year? [2]

(c) Outline the characteristics of the B horizon in:

a gleyed soil; a podzol; a tropical soil. [3]

(d) The pH of the A horizon in a podzol is acidic. What are the causes of this? [2]

How are the characteristics of the A horizon impodsols altered by farmers to make them suitable for cultivation? [2]

(e) Explain how human activity can change brown earths to podzols. [2]

(AEB) (AQA)

2 The diagram shows a simple model of nutrient re-cycling within an ecosystem.

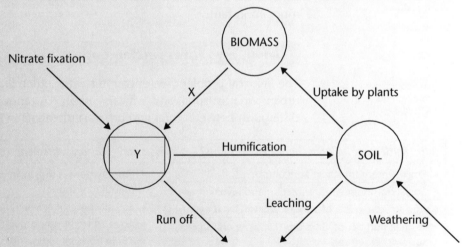

(a) On the diagram above, label flow X and storage Y. [2]

(b) Name **two** nutrients which are recycled in this way within ecosystems. [2]

(c) What is meant by the term biomass?

(d) Explain how nutrients are transferred from the soil to the plants. [2]

(e) Describe and explain three possible effects deforestation might have on soil. [2]

(OCR)

Chapter 6

Settlement issues

The following topics are covered in this chapter:

- Settlement distribution
- Settlement morphology
- Processes and problems of urbanisation

6.1 Settlement distribution

After studying this section you should be able to understand:

- the origins and growth of towns and villages, the importance of the rural area and how it is changing
- the hierarchical nature of settlement location
- models of settlement distribution

LEARNING SUMMARY

Settlement models and patterns

AQA A	U2	EDEXCEL A	U2
AQA B	UC	EDEXCEL B	U2
OCR A	U2	WJEC	U2
OCR B	U2	NICCEA	U2

Settlements are usually defined as 'a place in which people live and where they may be involved in various activities'.

> Quotations give colour to answers in exams.

Early settlement

> You could be asked to annotate a map with these features.

Before settling in an area, early farmers would have studied a range of factors relating to the **site** (the area upon which a settlement is built) and the **situation** (the relative location of a settlement to other features). These would have included the availability of clean water, freedom from flooding, level land for building and farming, timber for fuel and building, cultivable and grazeable soils, trade and commercial possibilities. Today, however, socio-economic factors are a good deal more important.

Urban and rural settlements

> This comparison is a common question at AS Level

The first real problem we encounter with settlement is the definition of rural and urban: what is 'rural' and what is 'urban'? A range of criteria have been used to distinguish between rural and urban settlements in the UK (see below).

Rural settlement (villages, hamlets and farms) – a comparison – Urban settlement (cities and towns)	
• Employment mostly in agriculture	• Employment is mostly in commerce, services and manufacturing
• Social homogeneity, interaction and involvement	
• Age structure: lots of 5 to 24 year olds and over 65s	• Socially heterogeneous
• A small number of functions (i.e. a general store, sub post office and a pub)	• Ages 24 to 54 predominate
	• There is a wide range (maybe hundreds) of functions (e.g. department store, banks, hospitals, etc.)
• Low density of population	
• Dispersed settlement	• High density of population
• Poorly developed infrastructure	• High density, closely packed housing.
• Census classifies the area as rural	• Biggest proportion of the population was born outside the urban centre
• There is land, as opposed to real money	
• Religion is important	• Limited responsibilities towards neighbours
• Inhabitants have a common purpose and community is generally 'inward' looking	• Well-developed infrastructure
	• The census classifies the area as urban
• Rural areas can be open, closed, disintegrated or integrated.	• Length of residence tends to be short
	• 'No' religion

KEY POINT

Many of the entries in the last table will apply equally well to other MEDCs and LEDCs, though variations between countries can be marked.

The problems of settlement classification and the dynamism of the settlements themselves are best exemplified through the study of a particular place. Urchfont, the example in the case study below, like many other small English settlements, continues to change in both its size and the functions it performs. The rapid speed of these changes – in this case from a small hamlet, to a large village to a commuter focus – further compounds the classification issue.

An English Village 1999: the idyll and the reality

Urchfont, in Wiltshire, is an exceptionally attractive village, but in this small (1100 people) settlement one can find social exclusion, agricultural decline, unsustainable development and environmental damage; but also hope for the future.

'Weekenders' (London is just 90 minutes away by car and train) are a problem for the village. Local people are being driven from the housing market, robbed of their homes! 7% of the houses in Urchfont are second homes, with a quarter of the village's population having arrived in the last 5 years. With house prices in Wiltshire rising 18% during 1999, first-time buyers can't afford the £100 000 price tag needed to buy a 'first home' in Urchfont.

Farming in the area is in decline; there are just a few remaining of the dozen or so farms that used to exist. Many farmhouses are now worth more than the acres of land around them! Many farmers have sold off land for redevelopment, but there is strong opposition from the countryside groups in Wiltshire to Urchfont's seemingly haphazard redevelopment.

Others see the 'invasion' more positively, as a community that is changing. The rapid growth of the village has seen the local primary school numbers increase and for the school to be revitalised. With a quarter of the population over 65, the influx of youngsters has brought balance back to the community. There is now a community bus, the skittles club has reformed and the pub is thriving.

In Urchfont you see in microcosm all the dilemmas of rural life.

CASE STUDY

Settlement hierarchies

Settlements can be classified according to size of population. There are no hard and fast rules but typical sizes might be:

Capital	5 000 000 or more
City	*c.* 1 000 000
Towns	10 000 to 100 000
Villages	500 to 2000
Hamlets	11 to 100
Isolated dwellings	1 to 10

decreasing size

Within a settlement, a range of activities take place that provide for the people who live there, e.g. it may have a hospital. This is a **function** (or service) the settlement performs.

Large settlements – cities – house many thousands of people and can support specialised goods and services. Smaller settlements support fewer people, and lower-order functions.

Models explaining settlement patterns

Models are used a lot in AS geography questions on the rural/urban world. Know the terminology.

Christaller: Central place theory (1933)

Uses To explain the size and spacing of settlements in MEDCs.

Assumptions That people who live in hamlets obtain low order goods locally, visiting higher order settlements to obtain goods not locally available. It's expected that high order settlements will be a minimal distance away. A hierarchy starts to reveal itself.

Results There appears to be a maximum distance people will travel for goods (range). There is a minimum number of people needed to keep a service in place (threshold). Every settlement has a sphere of influence (the area it serves). The above is described as Christaller's K=3 principle.

Central places

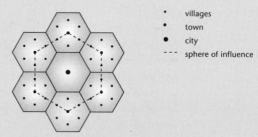

- ·   villages
- ·   town
- ●   city
- - - -   sphere of influence

OS maps are used commonly to describe city landuse distribution: know how to use this important skill. Exam questions may ask you to complete a graph or diagram. Christaller diagrams are ideal for this purpose.

Advantages and disadvantages It has been used in rural planning, is comparative and, on the flat plain it is designed for, works! Problems arise because people are not rational. Theory deals only with goods and services and chance plays a part in settlement location and pattern.

Losch: City rich and city poor (1941)

Uses To explain the patterns of settlement in MEDCs.

Assumptions That all settlements of the same size don't have the same functions. Sizes of central places vary with distance. Distinctive city rich – city poor pattern of sectors emerges.

Results Losch's theory appears to fit many of today's distributions, e.g. city poor equates to Aldershot and city rich to the Thames corridor, north of Aldershot, and the Reigate and Guildford areas to the south of Aldershot.

Advantages and disadvantages It is complex but does fit today's distributions.

Taafe, Morrill and Gould: Transport model (1970)

Uses Transport networks are used to develop and explain settlement patterns and emergence.

Assumptions That ports are built and penetrate into the hinterland. Nodes or smaller settlements develop and grow on the intersections and routeways out of the port.

Advantages and disadvantages It is dynamic, functional and explanatory.

KEY POINT

6.2 Settlement morphology

After studying this section you should be aware of:

- *the terminology for describing rural settlements*
- *the processes linking rural and urban developments*
- *the range of MEDC and LEDC models that help explain urban land use*

LEARNING SUMMARY

Rural patterns

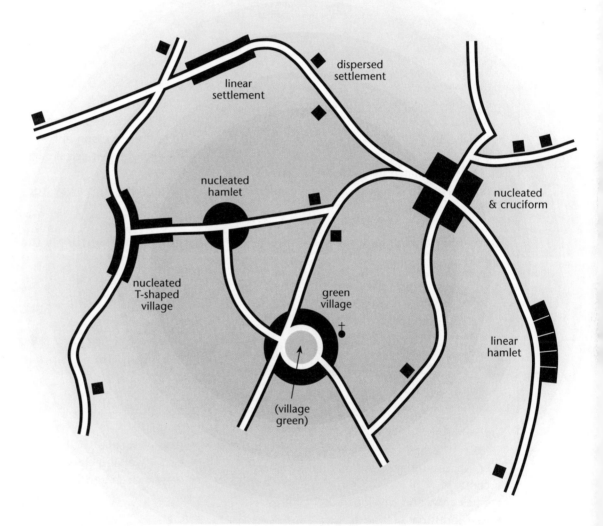

- **Nucleated** Houses and buildings are clustered. Traditional in the UK because of the enclosure system, defence and the need for water, e.g. Urchfont, Wiltshire.
- **Dispersed** Farms and buildings widely scattered. Common in sparsely populated areas, e.g. Oby, Norfolk.
- **Linear** Spread along a trade or transport route, e.g. Sutton Row, Norfolk.
- **Cruciform** Settlement occurs at the intersection of roads, e.g. Pottern, Wiltshire.
- **Green Village** A cluster of dwellings around a village green, e.g. Kimberley and Old Buckenham in Norfolk.

Urban models in MEDCs

These describe city layouts and are very simple.

The idea of 'bid rent theory' – that the value of land decreases away from the most central areas – underlies many of these models.

Burgess' concentric model (1920)

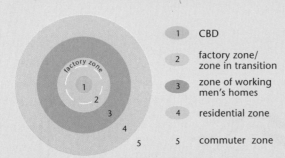

- 1 CBD
- 2 factory zone/ zone in transition
- 3 zone of working men's homes
- 4 residential zone
- 5 commuter zone

- city grows because of immigration and natural increase
- social class increases with distance from the CBD
- LBD is dominated by commercial activity
- population density peaks in the high-density, low-cost housing zone.

Hoyt's sector model (1939)

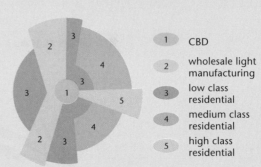

- 1 CBD
- 2 wholesale light manufacturing
- 3 low class residential
- 4 medium class residential
- 5 high class residential

- Hoyt emphasised the role of transport related to sector development
- he thought that certain activities deterred others
- better housing is away from industry.

Harris and Ullman's multiple nuclei model (1945)

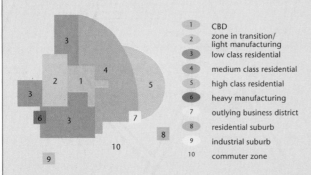

- 1 CBD
- 2 zone in transition/ light manufacturing
- 3 low class residential
- 4 medium class residential
- 5 high class residential
- 6 heavy manufacturing
- 7 outlying business district
- 8 residential suburb
- 9 industrial suburb
- 10 commuter zone

- emphasised that cities don't have a single centre
- cities grow and envelope other centres
- new industrial sites arise.

Mann's model land use in the UK city (1960)

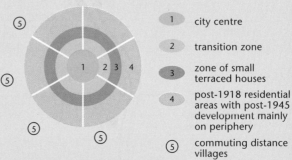

- 1 city centre
- 2 transition zone
- 3 zone of small terraced houses
- 4 post-1918 residential areas with post-1945 development mainly on periphery
- 5 commuting distance villages

- east–west split
- east = working class and industry
- west = clean and wealthy people's homes.

All these models are simplifications of reality. Cities are dynamic and unique.

KEY POINT

You need to know LEDC models as well as MEDC models.

Models of urban land use in LEDCs – Latin America

Many cities on the continent have the look of haphazard growth and a jumble of different activities and buildings without a clear structure. Large areas may be dedicated to residential land use without any commercial centres, industries may be located in central urban areas and airports may be within sprawling cities as expansion exceeds capacity to build infrastructures. High-class houses with all of the luxuries of excellent services may sit next to shanty towns with a glaring lack of even the most basic amenities.

Theories of urban growth

When looking at the models of urban growth in LEDCs you should always bear in mind the following points:

- the huge rate of urbanisation and in-migration
- lack of urban planning or finances
- the driving force of expansion is residential growth, with administrative, commercial and industrial coming a poor second.

Griffin and Ford Latin American concentric ring model

A concentric model similar to Burgess', with the oldest area in the centre – the old Colonial City. The growth of the cities occurred since the 1950s and was largely unplanned, and as most migrants could not afford to buy homes they engaged in self-build housing, e.g. *callampas* in Santiago de Chile, or *favelas* in Rio de Janeiro, and usually achieved by illegal occupation of land on the urban periphery. Exceptions occurred when there were sites available in the city centre such as steep hills, e.g. *morros* in Rio de Janeiro, riverbanks/flood plains, e.g. *Rio Bogota* in Bogota, or on building sites. Slowly shanty dwellings improved, with more permanent materials, the provision of roads and utilities, and became part of the city. Once this happened new areas of temporary accommodation developed further out from the centre and the process began again. Notice that there is little allowance for industry, recreation or commercial land uses.

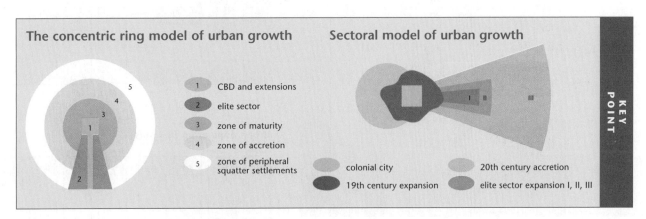

The concentric ring model of urban growth

1 CBD and extensions
2 elite sector
3 zone of maturity
4 zone of accretion
5 zone of peripheral squatter settlements

Sectoral model of urban growth

colonial city
19th century expansion
20th century accretion
elite sector expansion I, II, III

KEY POINT

Sectoral model of urban growth

This is a different kind of model, which relies on less rapid growth and strong physical restraints such as mountain topography, common to cities of the Andes along the Pacific Coast.

E.g. Caracas, Venezuela: sited in a narrow E–W valley with steep slopes prohibiting development; Bogota, Colombia: flanked to the E by steep mountain slopes and to the S by marshland.

The growth is linear rather than concentric; the outer expansion is not due to shantytown growth but to movement of the élite, occupying large areas at a low density. The poor serve to fill up rather than form the growth on the edge of the city.

6.3 Processes and problems of urbanisation

After studying this section you should understand:

- the causes and consequences of 19th-century growth on urban areas in MEDCs
- how rapid urbanisation has affected, more recently, the LEDCs and continues to affect the MEDCs
- the rise of the mega-city, primacy and the rank-size rule
- the changes that have occurred in the inner city areas of MEDCs and LEDCs
- that urban growth creates many problems and that much is being done to counter the problems
- that there is a need to manage the new wave of re-urbanisation

LEARNING SUMMARY

Urban development and urbanisation

AQA A	U2	EDEXCEL A	U2
OCR A	U2	EDEXCEL B	U2
OCR B	U2	WJEC	U2
		NICCEA	U2

Settlement questions at AS tend to be thematic and focus on a limited number of topics.

Urban development has changed the way that many of us live over the last century or so. In 1900, about 2% of population lived in urban areas, presently it is in the order of 50%+ (nearly 3 billion people), though clearly this percentage varies across the globe. It is suggested that in the past agricultural surpluses caused urban growth and development: today movement to the city is seen as a change in location, and lifestyles, as people are absorbed into a global society and economy.

Urban percentages for certain countries

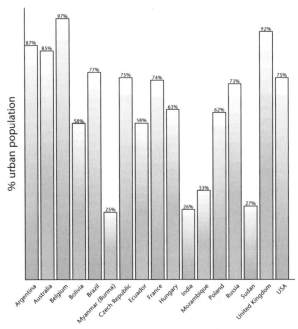

Source: Population Concern: 1995

Urban population trends and projections

Proportion of population living in urban centres

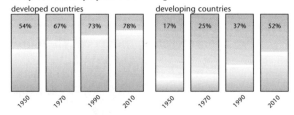

Urban population (millions)

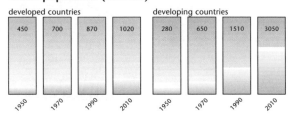

Source: United Nations. *World population prospects: 1990.* New York, 1991.

It's important you know why urbanisation has occurred. The reasons are similar in both MEDCs and LEDCs. There are clear differences between these 3 terms.

The urban world is today changing in three different but connected ways

- **Through urban growth** (due to population increases). World urban growth correlates with world population growth. Most of this growth is occurring in the LEDCs; for the most part MEDCs have static urban growth. In those LEDCs that are experiencing rapid urban growth they invariably still have a high degree of rural dwellers

- **through spreading urbanism** (the social and behavioural characteristics of those living an urban existence). This is usually obvious through shared activities and tastes. Urbanism or urban 'culture' has changed completely the views of many in the urban areas of Europe and the USA. The urban areas of most African and Asian cities have yet to taste this 'culture'.
- **and through urbanisation** (increases in the proportions of those living in towns and cities). This involves a shift in populations from rural to urban locations. Whether in LEDCs or MEDCs, urbanisation is seen as a cyclic process with populations moving from rural to industrial economies. There seems to be a balance of about 3:1/urban:rural at which growth flattens out. Urbanisation seems to have run its course in MEDCs with people returning to the countryside. Urbanisation is greatest in Asia and Africa at the moment.

It is generally agreed that urban growth and urbanisation over the last half century have been driven by the burgeoning global economy; the cities being the command and control centres of national, regional and, in some cases, global management of finance.

In Asia the global economy has almost single-handedly caused urban development. With transnational/multinational companies concentrating their factory production in countries and locations (i.e. near ports for export reasons) where there is a cheap and available labour pool, and where there may well be a growing home market for products. Such industrial provision makes the city look even more attractive than usual for the rural poor!

Urbanisation and urban growth in Britain

At the beginning of the 18th century, Britain was largely a rural farming economy. The largest city was London; other cities and towns, like Bristol and Norwich, acted as market centres for agricultural products.

Then notable changes in the agricultural landscape occurred, brought about by agricultural intensification leading to greater food production; driving many off the land, out of rural areas and into the cities in search of work.

The percentage living in urban areas in 1801 was 35%, by 1901 it was 78%.

The 19th century was a period of rapid industrialisation. Industry was located in the towns and cities and the increased job opportunities and higher wages attracted thousands to the city areas of the UK. Rapid rural to urban migration occurred: the young, innovative and energetic were attracted to the blossoming cities. By 1901 there were 33 cities of 100 000+, in 1801 there had been only one, London!

Cities in Victorian Britain became noisy, dirty, congested and dangerous places, packed with factories and plagued by disease.

People who could afford to, moved to the suburbs (**suburbanisation**) away from the filth and squalor. Aided by the development of the tram and railway networks the larger towns and cities of the UK began to spread beyond their original boundaries.

Urban areas have developed in the UK

- Through **isolation**, for effective government in the towns and boroughs of the UK.
- Through **interaction**, roads and rail bringing trade and people closer together.
- Through the **growth of conurbations**, following industrialisation.
- Through the **growth of an axial belt**, stretching between Merseyside, through the Midlands to London and the SE.
- **The metropolis develops**, with the continued growth of London and the SE.

Urbanisation in Latin America

Over the last 50 years or so the urban population of Latin America has risen exponentially: it has quite literally exploded. Once upon a time, Latin America was nearly totally rural in make up, now over 50% live in urban areas. The size, speed and scale of urbanisation has been massive on this continent.

City areas are not new to the continent; many date back to the time of the Aztecs and Incas, though most were initially established as ports during the colonial rules of European countries, for export, administration and commercial purposes. Most of the biggest urban areas are on Latin America's coast; on the whole, few people live in the interior.

> The rapid change and growth of cities in LEDCs make this a popular examination questions.

However, the biggest period of urban growth has been over the last 50 or 60 years, with growth rates between 5% and 2%. (At a conservative estimate, Mexico City sees at least 15 000 new faces per week.)

In most Latin American countries one city still dominates; though the single dominant primate city is becoming a thing of the past. Many of the middle rank cities are taking on the roles of government, both political and financial, as they continue to grow.

Some of the reasons for the rapid growth of the Latin American city are the same as that experienced in Europe, though for this area it has been compressed into a much smaller time frame (those reasons being migration and natural increase, at a time when the population was exploding!).

Mega-cities

Definition: 'a city with a population exceeding 8 000 000'

In 1950 there were just 2 of these, London and New York. Today there are at least 25 and, as 19 of them are in LEDCs, this section focuses on the developing-world city. Factors influencing mega-city growth include:

> ## The world's most populous cities

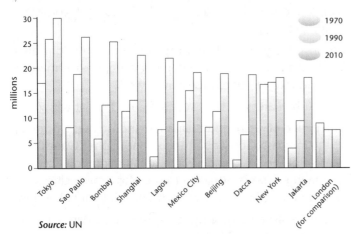

Source: UN

- rural to urban migration
- former ports continue to develop as trading sites
- industrialisation related to international production continues in huge (often transnational, foreign owned) sites; industrial technology transfer also attracts people
- decreasing mortality rates due to better medical provision resulting in natural population increase
- national development policies encourage or force people into urban areas.

Most mega-cities have slower growth now, than the millionaire cities (those with populations between 1 and 8 million). What is the challenge now is how to manage these sprawling settlements. It is a problem both for MEDCs and LEDCs.

> Urban primacy questions are a favourite at AS Level.

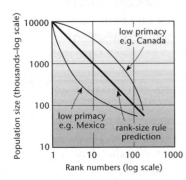

Primacy and the rank size rule.

The rank size rule, first drawn up in the 1940s, by Zipf, discovered an inverse relationship between the size and rank of a given settlement, i.e. the population of the second city in a country will be about half that of the biggest. This suggests that in any country there will be a few large places and many small ones. This means that when we investigate a country it will have high primacy if it exceeds the rank-size rule prediction and low primacy if it is less than the prediction. Generally rank-size patterns appeared in countries that are smaller than average, have only recently urbanised or have simple economic and social structures. Primacy is linked to development.

The rank-size rule does not explain settlement distributions, but does help to establish relationships between the size and importance of towns and cities.

CASE STUDY

The problems of mega-cities include:

- mega-cities become magnets for immigration
- shanty town and squatter settlements spontaneously appear
- hospitals and universities are located in the wealthier urban areas; rural areas have little or no provision
- informal employment grows but contributes little to the national economy
- poor air quality
- a lack of clean/potable water
- health problems associated with the above, e.g. asthma and cholera
- habitat loss
- crime and violence
- drug-related problems
- planning and administration are not co-ordinated

The advantages of mega-cities include:

- concentration of industry and finance ensures economies of scale
- general education and services are better in the big urban areas than they are in rural areas
- migrants deal with the housing problem in LEDCs!

As you can see the problems far outweigh the advantages.

KEY POINT

Urban problems

AQA A	U2	EDEXCEL A	U2
AQA B	UB	EDEXCEL B	U2
OCR A	U2	WJEC	U2
OCR B	U2	UEA	U2

Big cities have always suffered from the problems of overcrowding, crime, psychological stress, traffic chaos and pollution, whether in an LEDC or MEDC. The two places below are vastly different but growing cities, both experience the same sorts of problems.

Norwich's urban problems

1 Housing and urban sprawl
2 Water supply
3 Traffic
4 Homelessness
5 Prostitution
6 Heritage deprivation
7 Under-serviced and threat of out-of-town developments
8 Lack of open space
9 Industry and urban sprawl

Bogota's urban problems

1 Traffic
2 Pollution
3 Water and electricity
4 Political corruption
5 Housing and urban sprawl
6 Crime and safety
7 Earthquakes
8 Poverty/homelessness/prostitution
9 Industry and economic instability

Shanty towns

Use figures and data in your AS answers.

There are in excess of 250 000 000 people living in the shanty towns of the LEDCs. There are a range of causes and consequences. The main reason shanty towns exist is that the authorities just cannot keep pace with the influx of migrants to the city. Unable to live in permanent housing, people are forced to live in spontaneous settlements that they have built themselves, that use cheap waste materials, lack services (electricity, running water and rubbish collection), and are often sited in dangerous and vulnerable positions. They are overcrowded and harbour criminals. These conditions result in massive health problems, mostly related to dirty water, but compounded by poor diet. The shanty dwellers are the urban poor, the old rural poor, nothing changes for them. The young and better-educated are the migrants that 'make' it in the growing LEDC city. There are a range of problems that the authorities have to attend to to avoid shanty towns spreading uncontrollably and blighting the city.

Problems include:

- visual ugliness – they are an eyesore
- high incidence of disease and health problems in a large population
- fire hazard
- site hazards, most are on vulnerable sites, i.e. steep slopes, land that is regularly flooded or likely to collapse when earthquakes strike
- organised street crime and the drug trade use them as a base for trading and a convenient repository of goods
- house prices in nearby areas are severely depressed.

Many solutions have been suggested for the shanty problem, some of them very extreme, e.g. bulldozing happens periodically around the world. However, this is not the best 'cure' for the problem. It is better to confront it more positively, recognising that it is actually impossible to provide in quite the same way that shanties do for the new and the more established residents of these settlements.

Possible solutions include:

- making land available for new arrivals to the city
- making some tenure available, perhaps a ten-year lease
- making cheap material and waste materials available for building purposes
- introducing advisors into the 'towns' to help with construction advice
- encouraging a sense of community
- laying the foundations for 'self help' housing, and providing basic services, that new arrivals can build on.

The suburbs in the UK

By the end of the First World War London's urban area extended some 11 km from the centre, enabled by the growth of the railway. On the whole the suburbs were built around the villages that surrounded the city. The advent of the car speeded up the process, as did the relocation of industry to the suburbs. The imposition of green belt status has slowed the process but London is effectively surrounded by a suburban doughnut; stretching some 80+ km from the centre in places. Similar processes happen around all cities. In Norwich urban growth has swallowed up many of the suburban villages around the city, e.g. Catton, Eaton, Cringleford and Thorpe.

> How change affects the suburbs (and inner city) is a favourite at AS Level.

Characteristics of suburbs

- linear in nature
- low density, low storey housing with gardens and owner-occupied
- middle income families
- cul de sac, crescent and avenue layouts
- near modes of mass transport

Advantages of suburbs

- cheaper land prices
- modern houses with amenities
- pleasant, clean, pollution-free environment
- better schools?
- parks and gardens
- less crime

Disadvantages of suburbs

- increasing cost of housing
- lack of things to do in predominantly residential areas
- young people prefer the city, this changes the population structure
- difficulty of travel to the city: congestion and cost

Counterurbanisation

During the 70s and early 80s a noticeable process of counterurbanisation began in Europe and in the USA: a movement away from the city and into smaller, more community-based towns.

Reasons for this include:

- retirees moving to a more pleasant environment
- new towns growing outside the city, attracting workers
- decentralisation, due to high city rents
- the flight of city dwellers away from the congestion, stress and pollution of the city.

Urbanisation is now seen to pass through five important stages or phases:

1 people migrate from the countryside to the city
2 the pace of migration accelerates, suburbs grow
3 inner cities lose their populations
4 the whole city region loses population due to counterurbanisation
5 the population stabilises and people begin to return to the city – reurbanisation.

Problems and change in the UK city

The **Central Business District (CBD)** is generally at the heart of the city and is the focus of transport systems. The great accessibility of the CBD means that land is very expensive and limited in availability. CBDs display many common characteristics, see below.

Main features of central business districts

Concentration of shops	Concentration of offices	Little manufacturing industry	Growth of functional zones	Multi-storey development	Low residential population
Large department stores, such as Marks and Spencer, C&A and British Home Stores, are found at the heart of CBD. They attract large numbers of people from a wide area. Other specialist shops, such as book shops and jewellers, are also concentrated in the CBD.	Regional and head offices of large companies concentrate in the CBD. They are attracted by the accessibility of the city centre. Well-known companies like a well-known location for their head offices.	The CDB is not a suitable location for most manufacturing industries. However, a few specialized industries, such as newspaper and magazine publishers, do locate in the CBD. They need to be near to other CBD services and to have access to road and rail transport for distribution.	Similar activities tend to concentrate in certain parts of the CBD. It is usually possible to find areas given over almost entirely to entertainment, banks and financial services, educational facilities and shops.	The CBD has to grow upwards as well as outwards because of high land values. The most expensive sites have the tallest buildings. In a multi-storey block different activities may often occupy different floors.	There is little housing in the CBD because of the high land values. However, a few people live in luxury flats and apartments.

Source: Urban Landscapes, MacDonald Edre

Large cities often have recognisable landuse areas, i.e. financial areas, department store areas and other specialist areas.

The CBD changes constantly and continually to keep pace with a changing society. The **Core-Frame Model** attempts to reflect some of this dynamism. There are strong links between the core and frame, with areas and businesses constantly being assimilated and discarded from the core and frame. Elements of redevelopment, decentralisation, pedestrianisation, conservation and gentrification further complicate the pattern.

Core-frame model

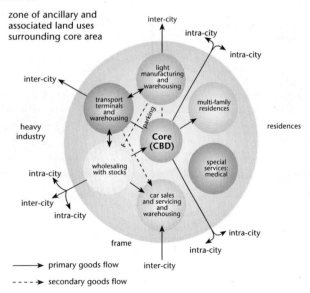

The changing pattern of urban functions

In the days before the car, shopping needs were satisfied locally on the street corner or on the high street. Increasingly location reflects a community's population, wealth and the infrastructure of the area. Spiralling town and city centre rents have forced offices out of the centre and into a range of cheaper and more advantageous sites around the city.

Retail locations

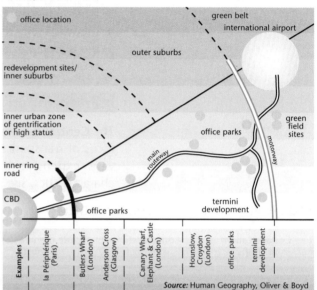

Source: Human Geography, Oliver & Boyd

Office locations

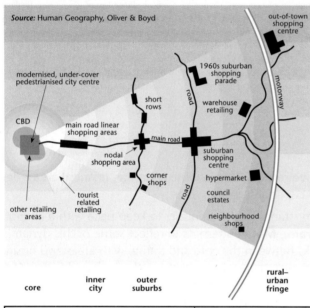

Source: Human Geography, Oliver & Boyd

As the largest land-user in the city, patterns of housing also reflect a more socially motivated variety of locations.

The cycle of residential land-use in the city

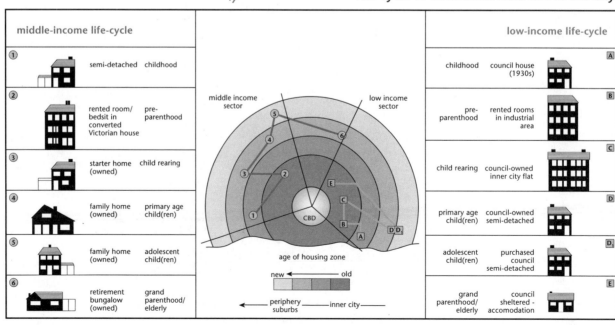

Beyond the CBD

The festering problems of the 4 000 000 that live in the inner city areas of the UK came to a head about twenty years ago with a number of serious riots, urban unrest and increased crime and racial attacks. This despite half a century of major government policy initiatives directed at inner city problems, starting with the 1947 Comprehensive Development Areas Policy (which spawned the New and Expanded Towns Act), the General Improvement and Urban Programme policies of the 60s, Community Programmes and Development Projects of the early 70s, the Urban Areas Act of the late 70s, through to Action for Cities and the formation of Urban Development Corporations, City Action Teams and the Inner City Initiatives in the late 80s. It was not until the late 80s that the problems caused by lack of decent housing, jobs and recreation facilities began to be dealt with. The range of problems that had to be dealt with are displayed below.

Problems in the inner city

Environmental problems	Economic problems	Social problems
• pollution • vandalism • dereliction • lack of open space • decaying housing • poorly built tower blocks • traffic congestion • social and educational services and recreation are poorly provided for • overcrowding	• unemployment • lack of skilled workers • poverty and low incomes • poor access • declining industry • lack of space for new industry • high land values	• crime problems • falling birth rates • concentrations of very old and young • endemic illness • lots of single parents • political activism • high concentrations of ethnic groups • dysfunctional families

KEY POINT

Despite all the changes that are now taking place, many people say that too little is being done to overcome the problems that continue to plague inner city areas. That said, several high-profile projects have moved the debate on, e.g. London's Docklands Project and the Cardiff Bay Development Project.

Managing cities

As urbanisation progresses many problems have to be tackled. Without proper planning and management problems, such as those emphasised in this chapter, can get out of hand. In the LEDCs governments and aid institutions have favoured big capital projects. But the real needs in the developing-world city may be more subtle: the creation of an environment in which jobs are available and small businesses can flourish; where land rights are recognised; where public transport and housing needs can be met at a reasonable cost; where child care, schooling, and health services are available; where law and order is seen to be fairly applied. The subtleties of the problem in the LEDCs are in many ways different to those experienced in the MEDCs. In MEDCs overcrowding, loss of agricultural land, urban sprawl and congestion, dereliction of inner cities and pollution are seen as more important, and this has spawned a raft of uniquely 'Western' solutions.

The green belt

This is land that is protected from industrial, transport and housing development and reserved for farming, forestry and wildlife areas.

There is a clear and obvious benefit from having and retaining green-belt land, but pressures for development, especially in the SE, will test governmental resolve to the full. Already the M25 has cut a swathe through the green belt around London and much more housing will have to be built on green-belt land to avoid the spoiling of the landscape beyond the zone of protection. It is estimated that the Channel Tunnel/Channel rail link and the planned development of Stansted Airport will create a need for 100 000s of homes on the green belt, to house those moving into the area. Further, without some development, such areas risk being spoilt by fly tipping, dereliction and general neglect.

New towns

The 1946 New Towns Act was put in place to provide homes for overspill populations from the biggest of the UK's inner cities, and to attract and develop new industrial agglomerations. They were meant to be self-contained communities separate from the cities. Most are near London, though successive legislation has distributed them over the whole of the UK. Many lessons have been learnt from the provision of these towns. They have met with varying success.

Cities of the future

Over the last century or so the city has emerged as the most popular type of settlement in which to live across the world. More people live in cities today than ever before. The future for most people looks to be an urban one.

Possibilities include:

- conflict city – where class and race hostility is rife
- the international city – a centre for trade, finance and high technology
- the neighbourhood city – emphasising the sense of community
- conservation city – where environmental concern is important, as is conservation of the historic
- leisure town – cities for the leisure age.

Whatever happens, it is likely that all settlements will be more sustainable: e.g. solar powered, with sewage digestion within the city, etc.

Sample question and model answer

1

(a) Outline the factors that help to explain the process of suburbanisation.   [15]

During the last century the industrial revolution created the need of a large workforce. This encouraged the process of urbanisation, and often the building of unplanned housing of poor quality.

Suburbanisation is the movement of people from the city centre out to the suburbs, and this occurred rapidly during and post-war as people sought a better standard of living and better transport made it possible to live further away from the workplace.

People moving to the suburbs were mainly middle-class families, as they could afford the larger houses with gardens, the expense of commuting to work, and thought that the suburbs were a safer and a cleaner environment.

Suburbanisation is still occurring. Populations of all large cities in the UK have decreased, for example London's population during 1970 to 1980 decreased by 3.1%.

> Remember to mention green belts and leapfrogging.

Industry also moved away from the centre's high land prices, and unattractive scenery, for cheaper, extensive land with room for expansion and an improved infrastructure, therefore attracting more skilled workers to the suburbs. This left the city centre with either young and unskilled inhabitants or elderly dependants, and an influx of educated families and professionals in the suburbs.

> 'Cottage' industries locating in suburbs also put prices up!

> Also retail parks!

An example of suburbanisation is Birmingham; this city encouraged low-density housing away from the city, compared to high-density housing in the city centre. These low-density areas are known as fringe belts and often have parks, golf courses and other leisure activities.

(b) What impact does suburbanisation have on socio-economic patterns within urban areas and the surrounding countryside?   [10]

> Impact on city.

The impact of suburbanisation on urban areas has often left a depleted population. The movement of people from the city centre left behind the unskilled, those unable to move, the elderly, those with insufficient funds to buy or rent suburban properties, and those who didn't drive or own cars who preferred to stay in the city.

> Spell this out: people with higher incomes can afford bigger, newer houses in more pleasant areas and can afford to commute to work either by public transport or by driving.

The movement of the professionals from the centre led to decreased employment opportunities and therefore unemployment resulting in increasing poverty.

The loss of the population also led to either vacant or unsightly derelict areas which were often thought of as eyesores and therefore decreasing the chance of future investments in the area. This can therefore be a vicious circle leading to a less attractive and unappealing city centre. But it can also lead to environmental improvement as derelict land can be cleared for recreational purposes.

Land also needs to be cleared to improve communications from the suburban to the urban area, such as buses and trains.

It can be seen that life within the inner city is of a poorer quality, the crime rate is higher, as is the unemployment rate, there are more people receiving housing benefit and there is also a higher death rate.

Sample question and model answer (continued)

'Necklace' villages may be absorbed into the suburbs if planning restraints are not upheld.

This answer shows an understanding of the topic.

It could be improved by:
- more examples
- mention of the slowing down of suburbanisation in MEDCs
- a little more on infiltration and succession by other groups later in the suburbanisation cycle
- including LEDCs – comments on shanty towns, the rich staying in the centre and the poor living around the periphery.

The city is therefore generally less well off as it loses money to the suburbs and outer-city industries.

Due to the expansion of the suburbs the land prices have increased along with demand, which has in turn put pressure on the amount of land, increasing the housing density and the pressure on the Green Belt. The pressure on land is also seen in local services such as schools, which raises the prices and can mean that locals may not be able to afford to remain in their village.

Also, due to the high percentage of commuters, villages are often left dormant, which can cause local shops or Post Offices for example, to close down. The commuters also increase the amount of traffic and air pollution.

(In the style of AQA)

Practice examination questions

One sort of question that will appear in a range of specifications is that which requires you to read a series of stimulus articles, data and/or tables. From this information extended prose will be sought in response to a range of focused questions.

1 Read these articles thoroughly.

We must start planning for an urban future

We don't, in Britain, think much about cities.

Nonetheless, there are UK cities that think big. Glasgow, the former European City of Culture, has come to define itself as a mecca for arts tourists. Birmingham and Manchester celebrate their multiple personalities as regional capitals and, increasingly, European metropoles.

Thanks to Symphony Hall, parts of central Birmingham have become newly desirable – for that hard-to-define population of younger people and the pizza-deliverers and club-owners who cater to them. Flats are being bought and rented; housing associations and private developers are sizing up opportunities. The flight from the centre starts to reverse itself. If there is a lesson, it is not about 'planning' in the sense of a government man with a map. It is about the creation of opportunities – commercial and cultural, intertwined – that have a prospect of being realised in their natural urban environment.

There is a delightful serendipity in the way that cultural development spins off economic and housing development. That is not at all the same as saying, leave it to the market. Markets don't normally build symphony orchestras. What has been happening in Birmingham owes a lot to the determination and consistency of the city council. Public money matters. The imagination of city leaders (public and private) also matters.

What Birmingham's example says is that we need to revise our conception of planning. What government can do is lay down frameworks within which market-led development can take its own course; private interest can be bent to public purpose. This is the model of planning needed as we start to think about where housing is going to go to contain the huge growth in household numbers projected by the government.

However, we cannot forecast precisely how society will adapt to changed expectations, nonetheless household demand is rising and space will have to be found for more homes. In some quarters there is talk of new garden cities and huge infrastructure schemes. But they would require the rebirth of Big Government, and where is the mandate for that? Stevenage and Crawley and most of the other New Towns worked. Government showed itself able to create value by developing empty land and selling it, to householders and industry. But that was then. Government now has to go with the flow of private development.

One government duty is to ensure that the supply of housing affordable by those on lower incomes is kept up. That means making it easier for private landlords to let while subsidising social landlords. Imaginative local authorities have cut deals with developers that reserve land for housing associations. Government financial rules should make these easier not harder.

Planning is essentially a local matter. It is for local authorities to zone and developers and their customers to identify sites for building. Central government is the court of appeal. If planning pressures are going to grow, Mr Gummer and his successor would be well advised to streamline the process of inspection and final judgement.

Also, there is a proper national concern for greenery, in the form of green belts and the like.

Adapted from *The Independent* 6.6.96

Where can we put 4 million homes?

A map published by the Department of the Environment pinpoints the likely pressure points in the next 20 years.

The biggest increase is expected in London and around Manchester and Liverpool, though here derelict land can be used.

Hampshire, Kent, Essex, Buckinghamshire and Bedfordshire are all earmarked for up to 160,000 new houses; Cheshire, Gloucestershire, Wiltshire, Berkshire, Norfolk and Devon may need between 80,000 and 120,000.

Inevitably, say planners, this will mean expanding existing towns on to green field sites. It may even require taking a slice of green belt land, or the construction of new towns.

The time scale for finding space for the homes, according to officials, is three years. If some councils have failed to provide sufficient room for extra housing by the end of the decade the Government can insist they do.

On a regional basis, an area of rural land larger than Greater London is at risk, the CPRE says.

But the Department of the Environment says that, nationwide, the plans would mean only 1.3 per cent of extra urban land.

Millions of new houses will have to be built across the English countryside over the next 20 years to cater for the break-up of the traditional family and the growing number of people living longer, said John Gummer, the Environment Secretary yesterday.

In a keynote speech to the Royal Town Planning Institute's annual conference in Brighton, Mr Gummer sought to open a national debate on one of the most pressing environmental and political issues confronting the Government: the need to accommodate an estimated 4.4 million extra households by 2016.

Mr Gummer said he did not want to pre-empt decisions over where the new houses should be built, but neither was he ruling out any options – from the establishment of free-standing new settlements to the expansion of existing market towns.

According to Environment Department figures, the shires will face the biggest growth in development once the reclamation of derelict city land has reached saturation point.

Mr Gummer said he did not have any ready-made answers to the questions he posed, but he added: 'Unless we are all prepared to discuss the issues frankly, openly and fully, we will pay the environmental and social price of these new lifestyles without ever having decided whether we were prepared to afford it.'

The Council for the Protection of Rural England called for an overhaul of the existing planning system to avoid scarring the countryside with development.

Tony Burton, head of planning, said: 'Rural areas are already under serious threat from the Government's existing plans for new housing. These levels are set to rise if the latest projections are allowed to dominate future plans at the expense of the environment.'

Mr Burton said the figures helped to define the problem the Government faced but should not be regarded as a target.

Roger Humber, director of the Housebuilders' Federation, said the issue of building new settlements would have to be faced because there was a limit to how far existing towns could be 'crammed'.

But he said Mr Gummer, or any successor at the DoE, had a difficult task planning new development since ministerial powers had been limited by Parliament. Local authorities now certify their own structure plans and many are fiercely resisting pressure to increase the number of new houses they plan to build.

'Sooner or later, legislation will be needed to ensure there is a strategic element to all of this,' said Mr Humber. 'At the moment, it is too tactical and local an issue.'

Professor Cliff Hague, president of the Royal Town Planning Institute, also said strategic decisions were needed to avert 'a war' between cities and the countryside, builders and environmentalists.

Adapted from The Times 6.6.96

Urban sprawl

According to the new research by the Council for the Protection of Rural England, oases of peace have shrunk over the past 30 years to only three major areas of rural tranquillity.

In new maps issued by the CPRE, the three large remaining 'reservoirs' of tranquillity are in north Devon, the Marches of Shropshire and Herefordshire, and the Pennines.

The parts of England which remain free from urban blight, noise, pollution and over-crowding are estimated to be four times smaller than in the Sixties. Over the past 30 years an area of tranquillity the size of Wales has been lost. The CPRE says industrial development, new roads and increasing traffic have left the countryside 'shattered'.

The research is the most recent attempt to measure 'quality of life' factors once regarded as unquantifiable, but 'environmental evaluation' is now acknowledged as important enough for the Department of Transport to carry out a noise and air pollution study.

Tranquil areas are a planning tool developed for the CPRE by the ASII consulting group to replace the simple split between rural and urban areas regarded as too narrow.

Tony Burton, the CPRE's senior planner, said: 'The built-up areas of England represent 17 per cent, but the disturbed areas represent 50 per cent. It is crucial to acknowledge that development goes far beyond the physical boundaries of brick and concrete.'

'Tranquil areas are defined as places beyond the immediate influence of towns, roads, airports, overhead pylons and mining.

Examining the shrinkage over the past 30 years shows, for example, that in the Southeast, the reasonably intact countryside surrounding London in 1960 has completely fragmented. The CPRE blames a four-fold increase in road traffic, air traffic, new developments and the rise of the National Grid. It estimates it requires 150 miles travel from the capital to reach an area of tranquillity.

The Southwest, formerly England's most tranquil area, is breaking up fastest; East Anglia is 'fragmenting'; while the North east is the least changed since the 1960s.

Practice examination questions (continued)

Regional changes in tranquil areas over three decades

	1960s	*1990s*	*Change*
East Anglia	72%	64%	– 8%
East Midlands	70%	56%	–14%
North East	75%	68%	– 7%
North West	67%	55%	–12%
South East	58%	38%	–20%
South West	83%	66%	–17%
West Midlands	70%	55%	–15%
Yorkshire & Humberside	74%	60%	–14%
ENGLAND	70%	56%	–14%

Tranquil areas are measured as...

- 4km from a power station
- 3km from major motorways, large towns and major industrial areas
- 2km from major trunk roads and the edge of smaller towns
- 1km from busy roads and the main-line railway lines
- beyond military and civilian airfield noise
- beyond sight of open-cast mining

Adapted from *The Independent* 1.12.95

Where will we live in the 21st century?

The 19th century gave us the terrace. The 20th century gave us the semi. What will be the 21st-century equivalent?

A planner and an economist from the Urban and Economic Development Group (URBED), London-based consultants, have for the past two years had their thinking hats on. The result is *21st Century Homes: Building to Last*, sponsored by the Joseph Rowntree Foundation, the housing charity.

Pollution from cars means that we should not be building houses which require people to drive.

The answer is to build at higher densities in larger communities which can support local public transport, shops and schools. 'New houses should be built within existing settlements to reduce commuting.' We must develop models for 'sustainable urban neighbourhoods' which will be as attractive to new families as today's suburbs.

The government has already enshrined this principle in its planning guidance, which states that new housing should be built within existing towns. If this is not possible, then developments should be sufficiently large, by reaching a population of about 20,000 in 20 years, to become free-standing with a degree of self-sufficiency.

There are other environmental needs, too. The Government aims to reduce CO_2 emissions to 1990 levels by 2000, and energy saving in domestic heating is a key element. We should be building thermally-efficient homes. New homes can become greener at minimal extra cost by avoiding CFCs, reducing water consumption, providing facilities for waste recycling, avoiding harmful materials and ensuring effective ventilation.

A second set of needs is demographic and social. The Government predicts that there will be four million new households by 2012, and few of these will conform to traditional models. The nuclear family now accounts for one in four house-holds. Thirty per cent of households rely on pensions, and, most striking, 85 per cent of the increase in households between 1989 and 2011 will be among single people.

It is essential to foster a sense of community in new housing. However much one introduces principles of being 'secure by design', aimed at reducing crime by encouraging residents to watch over their own streets, such schemes will fail unless residents feel part of a community.

More flexible homes would help. 'A new home should be able to change to demographic demands, so that it can serve as a home for life. We must have a greater range of housing types, as well as increased internal flexibility.' More flats are needed in city centres. Layouts should be open-plan, with flexible arrangement of bedrooms.

The report gives three examples of where these principles have been put into practice. One scheme was in Milton Keynes, where a terrace of five houses was built with moveable internal walls. Another was in Manchester, on the Hulme Estate, where residents were involved in the design, and homes were linked to jobs, in a scheme for 50 single people. In Swansea, the Gwalia Housing Society redeveloped its estate to be more energy-efficient and secure.

The pitfalls were many. Home-owners fear innovation. They are confused by environmental designers with conflicting views. Contractors increased the cost of tenders in response to innovations. The conventional way of valuing houses, by developers and buyers alike, fails to recognise energy efficiency, stresses the number of bedrooms rather than the floor area and values conventional design. Housing associations the innovators in house types, are under pressure to produce more, rather than better, homes.

As usual, the future lies with government, central and local, which must encourage more positive planning. But it also lies with the consumer. 'Consumers must press for change.'

How one community has planned for its future

Part of the City Challenge redevelopment of the Hulme Estate, Manchester, is the Homes for Change project. A mixed-use £2.7-million development of 50 flats and 15,000 sq ft of work-space is being developed with the Guinness Trust housing association.

Plans centre on an urban block, rising to six storeys and topped by grass roof-gardens around a central courtyard, on the site of the former Naval Brewery on Chichester Road.

The project is committed to an environmentally friendly design. A specialist technical architect included targets for building with sustainable materials, reduced CFCs and reduced heating costs to cut CO_2 emissions. The group is made up of those who used to live in council flats in Hulme. The studios, shops, café, office space and small theatre should provide a heart to the community.

Adapted from *The Times* 6.9.95

Section A

1 Read the articles thoroughly then answer the following questions.

(a) (i) Outline why we need to start planning for our urban future. [8]

(ii) Explain what is wrong with our current planning system, how can its weaknesses be overcome? [4]

(b) (i) Comment on the national concern for green belt retention. [8]

(ii) Imagine you are a politician: outline and justify your reasons for building on the Green Belt and not leap-frogging beyond it. [4]

(c) (i) Outline the reasons why urban dwellers still seek refuge in the rural areas. [4]

(ii) 'Urban sprawl has to be controlled'. Comment on this statement. [8]

(d) 'Where will we live in the 21st Century?' Comment on the environmental, social and demographic factors that will need to be considered as plans are drawn up for the 21st Century. [12]

Section B

2 (a) To what extent does physical geography contribute to the siting and growth of rural settlements? [10]

(b) In what ways and for what reasons have rural settlements changed? [15]

AQA

Chapter 7

The dynamics of population

The following topics are covered in this chapter:

- *Population growth and distribution*
- *Population structure*
- *Population movements*
- *Population control*

7.1 Population growth and distribution

LEARNING SUMMARY

After studying this section you should understand:

- *why and how populations have grown and changed*
- *the concept of optimum population*
- *a population density and distribution*

In 1798, the Rev. Thomas Malthus wrote his *Essay on Population*, in which he warned that population would outstrip food supply. Comparatively recently, Paul Ehrlich in the *Population Bomb* (1970) warned of impending disaster if the world's population was not brought under control. The growth of the world's population dominates governments, planners and world history. And it is where we start this chapter.

> World population in 2000 = 6 billion.

World population change

AQA A	U2	EDEXCEL A	U2
AQA B	U2	EDEXCEL B	U2
OCR A	U2	WJEC	U2
OCR B	U2	NICCEA	U2

About a million or so years ago the first humans walked the earth; for thousands of years population changed little, probably reaching 250 000 000 at the time of the Christian era. Change continued to remain low until about 200 years ago, linked initially to the industrial revolution in Europe. From this point the characteristics are:

- World population reached 1000 million in the early 19th century, 2000 million in the 1920s, 3000 million in the 1960s, and so on. An extra 1000 million added in 100, 40 and 10 years, respectively.
- World population doubled between 1820 and 1920. It doubled again in 50 years up to the 1970s. Theoretically it will double again in the next 30 years.

> Important at AS as it's so dynamic.

- The 'worlds population is increasing at an increasing rate'.

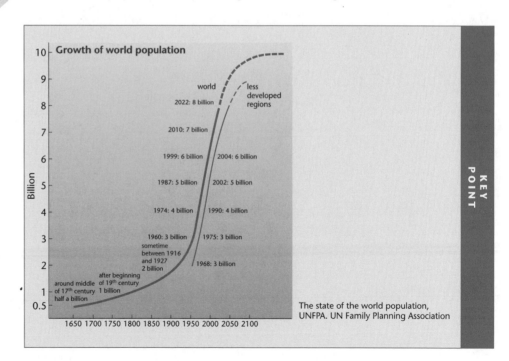

The state of the world population, UNFPA. UN Family Planning Association

Population has not increased evenly over the world: some countries, especially in Africa and Latin America since 1950, have grown exponentially. Asia has during all of this period of time maintained a very high half of the world's population! The growth rates of some North African countries is in the order of 3% (enough to double the population in 25 years). This compares with many European countries who have experienced drops in population in recent years. It is estimated that 95% of growth will occur in Africa, Asia and Latin America over the next 25 years. That birth rates may continue to increase over the next 50 years is assured, the parents have already been born!

Causes of change

If the number of births (birth rate = BR) exceeds the number of deaths (death rate = DR) in any one year then populations tend to increase. This is called Natural Increase. In addition, if the number of immigrants that join a country exceeds emigration, populations will rise.

> **KEY POINT**
>
> Three factors influence population change:
>
> Fertility +/– Mortality +/– Migration = +/– Growth

Reasons for change

The reasons for falling death rates

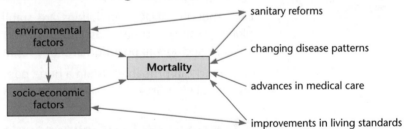

The influence of falling death rates

It's important to understand all of this: it's good essay material.

Through the 20th century, Africa has seen death rates halved, particularly amongst children. This has been brought about with improvements in sanitation and improved medical techniques, medicines and general provision. Crude Death Rate (the number of deaths/1000 inhabitants/year) is the most common measure of mortality. However, as it takes no account of population structure it is of little use (a better measure is infant mortality, the number of deaths of infants under the age of one year old/1000 live births). The northern hemisphere generally has the lower DRs.

The influence of birth rates

BRs in the MEDCs have fallen, in the LEDCs high BRs have persisted but are falling in some. This will, as outlined above, result in these countries' populations doubling over the next half century. At the same time, MEDCs' populations will shrink.

Why birth rates decline in MEDCs

Reasons include:

- availability of family planning
- increased education and literacy
- better health and fewer child deaths
- more employment opportunities
- later marriage
- migration to the cities
- better deals for women
- more income and rising living standards.

Why LEDCs continue to have high birth rates

The reasons for the continuing high BR in LEDCs (the number of live births/1000 people/year) is a complex problem to untangle. Reasons include:

- the importance placed upon child-bearing in some countries
- large families are seen as insurance for the future
- women are disempowered
- more children = more workers.

Generally population increase equates to high BRs (e.g. in Africa the BR is in excess of 40/1000; in Europe it's less than 15/1000!).

Implications of increased population

If population increases the following increases will also occur:

- fuel consumption
- greenhouse effects, e.g. flooding, salinisation, rising sea levels
- acid rain
- waste and pollution
- agricultural activity, to feed the growing world population, causing greater use of pesticides and fertilisers
- use, and therefore depletion, of world minerals, hardwoods, etc.

	MEDCs	LEDCs
% of population	25	75
% resource consumption	80	20

The ability of the earth to sustain human life is limited – resources can only be exploited to a certain level before they are exhausted – a limit known as the **carrying capacity**.

Even though MEDC populations are stable, they are the biggest consumers of global resources, though as LEDCs strive to industrialise their consumption will increase and so will pollution and environmental decay.

The world can probably sustain a huge population for a short time, but medium to low population growth rate would enable a more sustainable future.

The concept of optimum population

This concept involves an ideal population living and working in a given area: numbers of people in balance with resources maximising GNP.

- If population is > than resources, it is said to be **overpopulated**.
- If population is < than resources, it is **underpopulated**.

The best AS students will provide accurate definitions of optimum, over- and under-population.

Though the idea of an optimum population is an interesting one, it is the extremes, overpopulation and underpopulation, that draw most attention today.

Underpopulation

Causes

- physical disadvantage, e.g. climate
- inaccessible/poor communications/remote
- historical, e.g. Australia
- types of economy, i.e. intensive manufacturing and/or agricultural
- small indigenous population, e.g. Brazil, 92% live in the SE and there are lots of natural resources as yet untouched; Canada, wealth based on fish and forestry.

Underpopulation does not imply a country is poor or has a low population density.

Consequences

- resources developed by foreign countries
- regional disparity is obvious
- high urbanisation
- high standard of living
- high immigration.

KEY POINT

Overpopulation

The best students will be well aware of these views on under- and over-population, and have good case studies to support them.

Causes

With regard to over population there are two opposing views:

- The Neo-Malthusian approach, that 'increasing population leads to environmental degradation, which limits population growth'.
- The Boserupian approach (after Ester Boserup, a Danish Economist), that 'necessity is the mother of invention. Increasing population drives agricultural productivity, which allows further increases in population'!

Consequences

- starvation
- malnutrition
- poor health
- lack of jobs
- slow economic growth, e.g. India and Egypt.

KEY POINT

Population density and distribution

AQA A	U2	OCR A	U2
EDEXCEL A	U2	OCR B	U2
EDEXCEL B	U2	WJEC	U2
		NICCEA	U2

Density

A measure of the average number of people/unit area. This measure should not be used to compare countries in terms of overpopulation as different countries have different carrying capacities.

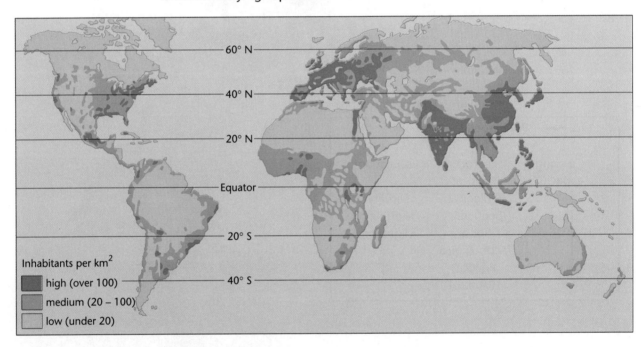

Inhabitants per km^2
- high (over 100)
- medium (20 – 100)
- low (under 20)

Distribution

Relates to location based mainly on economic and physical factors. It is difficult to measure as it is a spatial indicator.

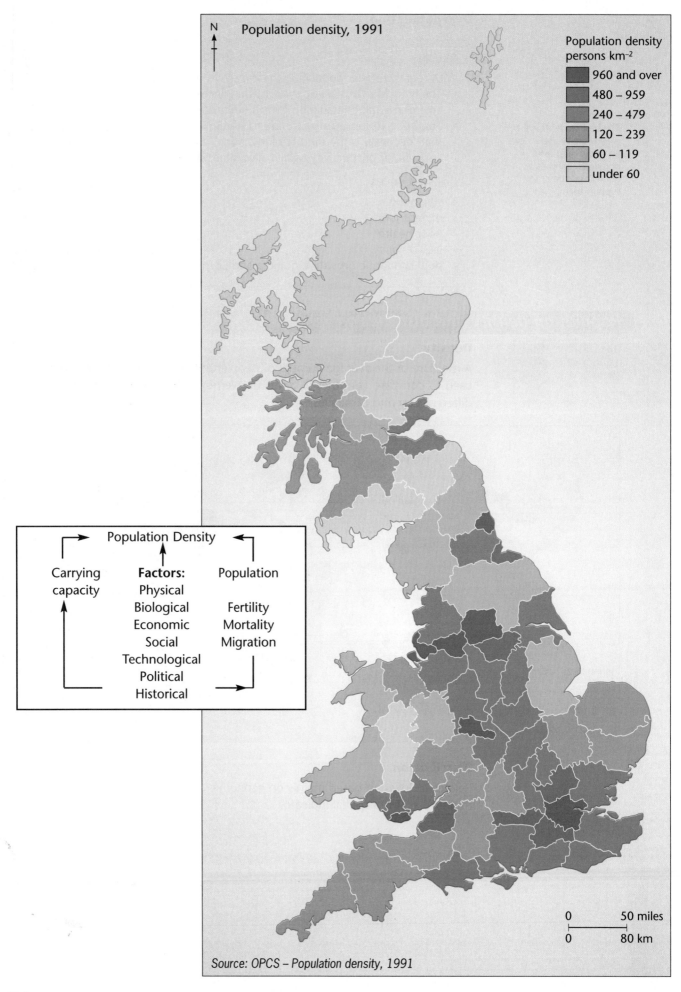

Population density, 1991

Population density
persons km⁻²

960 and over
480 – 959
240 – 479
120 – 239
60 – 119
under 60

Population Density

Carrying
capacity

Factors:
Physical
Biological
Economic
Social
Technological
Political
Historical

Population

Fertility
Mortality
Migration

0   50 miles
0   80 km

Source: OPCS – Population density, 1991

7.2 Population structure

After studying this section you should understand:

- *the use of population pyramids to analyse age and gender structures*
- *how these can be used to study the proportion of the youthful, mature and elderly populations*
- *how the Demographic Transition Model uses population structures to assess development*

Population pyramids

AQA A	U2	WJEC	U2
OCR A	U2	NICCEA	U2
OCR B	U2		

Population pyramids are a graphical method of representing the age and sex structure of a population at one point in time. The shape of a country's pyramid is the result of past fertility, mortality and migration within the population. It is sensitive to 'baby booms', wars, epidemics, population planning policies and so on.

Learn these shapes and be able to apply them to any given country.

On the pyramid each group is represented by percentages, therefore comparisons in the age–sex ratios can be made between countries.

Three main types of population pyramid can be distinguished:

- stationary
- progressive
- regressive.

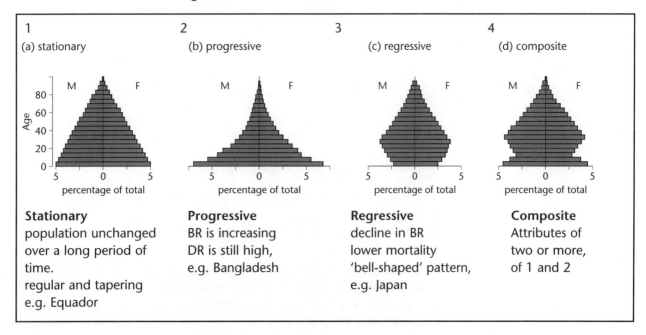

1
(a) stationary

Stationary
population unchanged over a long period of time.
regular and tapering
e.g. Equador

2
(b) progressive

Progressive
BR is increasing
DR is still high,
e.g. Bangladesh

3
(c) regressive

Regressive
decline in BR
lower mortality
'bell-shaped' pattern,
e.g. Japan

4
(d) composite

Composite
Attributes of
two or more,
of 1 and 2

Dependency

One of the most common measures derived from population pyramids is the **Dependency Ratio**. For convenience, the age structure is divided into three broad age bands:

- youthful dependants aged 1 to 14
- elderly dependants aged 65 and over
- working population aged 15 to 64.

Current problem, therefore important at AS!

To calculate the proportions of dependants, divide the proportions of young and old by the proportion who are economically active, to yield a ratio.

If the dependent group constitutes a relatively high proportion, the ratio will be high. The greater the proportion in the working age group the fewer the dependants. Dependency ratios are higher for developing countries.

Pyramids for actual countries may combine attributes of different types. These are termed composite.

Points to bear in mind:

- Is a young dependant more of a burden on a working person's resources than an elderly dependant?
- Many over 15s continue to study so this age group is now incorrectly classified in the working population.
- In many LEDCs children start jobs at young ages.
- How should students, housewives and the unemployed be classified?

Trends in population structure in LEDCs and MEDCs

LEDCs	MEDCs
General	*General*
• BR is still high, but reducing • the numbers of those who are 15 or less is still very high • there are more old people	• increased numbers of elderly • the upper end of the population structure is increasingly fit and healthy • reduced BRs • decline in the working population
General effects on the population structure	*General effects on the population structure*
• population will continue to increase; anti-natal population policies will be imposed to deal with increased population • pressure on the countries' economies • children need to work • migration increases • unrest and instability within the country	• tax burden on workers increases • greater Government spending • fit OAPs retire later • career paths close because of the above • the state is unable to fully provide for OAPs • the elderly are moneyed and mobile • growth in 'grey' investment/economy, holidays and retirement homes

KEY POINT

The 'greying' world population

As countries become 'developed' they experience a static population with low BRs and DRs. It is the lowering of the DR that is most important as, by 2050 1:7 of the world's population will be over 60. The world's population is 'greying' fast and this affects all of us.

- **Supporting the elderly** In the MEDCs increasingly the old will have to look after themselves in terms of pension provision.
- **Impact on young** The 'greying effect' impacts on the young, in many countries as it leads to unstable structures and economies. Clearly the situation differs in LEDCs and MEDCs. Compare, for instance, the UK, Japan and Kenya!

The Demographic Transition Model (DTM)

AQA A	U2	EDEXCEL A	U2
OCR A	U2	WJEC	U2
OCR B	U2	NICCEA	U2

The differences between the two ratios of **Crude Birth Rate** (CBR) and **Crude Death Rate** (CDR) are known as the rate of **Natural Increase**. Utilising these two simple indices it is possible to analyse the idea of demographic transition and change in the rate of natural increase over time.

The demographic transition model

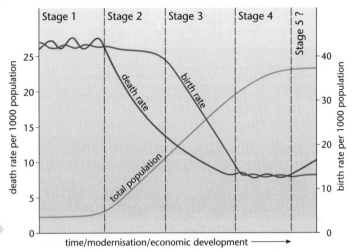

Adored by examiners. A favourite essay topic – always use examples.

	STAGE 1	STAGE 2	STAGE 3	STAGE 4
	DR high – little medical care BR high – no birth control and children an economic advantage DR fluctuates due to plagues/famines	DR declines – medical developments improved nutrition and sanitation BR remains high – children remain an economic advantage since urbanisation and mechanisation at an early stage, and births seen as desirable Increasing difference between BR and DR	DR low and slowly decreasing – continued medical and nutritional developments BR starts to decrease rapidly – improved education and availability of contraception and decreased economic value of children due to increasing urbanisation and use of technology Decreasing difference between BR and DR	DR remains low and slowly falling – continued medical progress and enhanced welfare provision BR declines to just above DR – economic independence of women, improved contraception and changing views on desirability of births in highly urbanised society
	high proportion are young	very high proportion are young	increasing numbers surviving to old age	high proportion of population are ageing
	Stage 1: high stationary *Pre-industrial society*	Stage 2: early expanding *Early industrialisation*	Stage 3: late expanding *Later industrialisation*	Stage 4: low stationary *Developed country*

DTM is used widely to predict future population patterns and numbers.

Always look for anomalies in a county's DTM fit or consider data.

The validity of the DTM

DTM and Europe	DTM and LEDCs
• Suggests fertility decline is even. Countries in Europe do vary! • Suggests fertility decline be linked to increases in literacy, urbanisation and decreases in mortality. • Ignores birth control in stage 1. • Delays in child bearing ignored in later stages. • Does the fertility of one generation affect the next? • What effect pro-natalism, fascism and Catholicism?	• The transitions in mortality and fertility occur in a shorter time. • Growth in population is greater in LEDCs. • Comparisons with MEDCs, because of the above, are difficult. • DRs have fallen faster, and for different reasons to Europe. • Mass contraception availability important.
General uses	**General limitations**
• As a descriptor, for comparison, for prediction, for explanation, as a starting point for discussion. • It does not predict or assess when, or how long transitions will be.	• Limited database. • Migration is not assessed. • External influences not considered. • Does not go beyond Stage 4.

Applying the DTM

To the UK

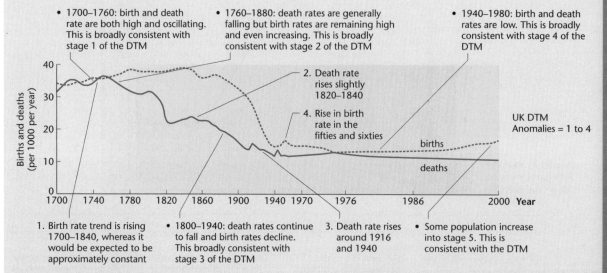

- 1700–1760: birth and death rate are both high and oscillating. This is broadly consistent with stage 1 of the DTM

- 1760–1880: death rates are generally falling but birth rates are remaining high and even increasing. This is broadly consistent with stage 2 of the DTM

- 1940–1980: birth and death rates are low. This is broadly consistent with stage 4 of the DTM

2. Death rate rises slightly 1820–1840

4. Rise in birth rate in the fifties and sixties

UK DTM
Anomalies = 1 to 4

births

deaths

1. Birth rate trend is rising 1700–1840, whereas it would be expected to be approximately constant

- 1800–1940: death rates continue to fall and birth rates decline. This broadly consistent with stage 3 of the DTM

3. Death rate rises around 1916 and 1940

- Some population increase into stage 5. This is consistent with the DTM

Why did mortality decline?

- improved sanitation and hygiene
- improved food supply
- reduction in disease impact
- medical advance
- rise in living standards.

Why did fertility decline?

- legislation to do with female and child labour
- more women in the work force
- contraception
- declining infant mortality.

To LEDCs

Most are bunched in the second stage, unable to achieve economic and social progress to enable them to move on. Population increases and ecological problems further hamper progress. However, there have been notable decreases in mortality and fertility in the last 100 years.

Why has mortality decreased?

- malaria and other tropical diseases have been eradicated or controlled
- improved health care
- stronger economies
- better nutrition.

Why has fertility decreased?

- age of marriage has increased
- contraception
- urbanisation/Western values have been taken on
- the status of women has improved.

7.3 Population movements

After studying this section you should understand:

- *the causes and consequences of migration and its profound effects*
- *types of migration*
- *models used to describe migration patterns*

LEARNING SUMMARY

Migration

AQA A	U2	EDEXCEL A	U2
OCR A	U2	WJEC	U2, A2
OCR B	U2	NICCEA	U2, A2

Technology and economic progress increase mobility and permit increased migration. Migration is usually defined as a 'change of residence of substantial duration' (a year or more). Migration can be classified by the following.

Timescale

- seasonal in nature, e.g. Mexican fruit pickers in California
- temporary, e.g. asylum seekers
- periodic, e.g. forced out by conflict
- permanent (over 1+ years), e.g. emigration to New Zealand.

Distance

- internal – within cities
- external – a move abroad
- inter-regional – a job move
- international – emigration/immigration.

Causation

- forced, e.g. Kosovan refugees
- spontaneous
- free } could be politically, socially or economically motivated.
- planned

> Continually in the news. Keep your 'scrapbook' up to date! Good case studies appear just about every day, therefore few here!

Migration does tend to be highly selective, but most migrants tend to have a combination of traits.

- **Age** The majority of migrants are between 18 and 35, often moving for a first job. Increasingly people migrate on retirement.
- **Sex** In general in MEDCs male and females migrate in roughly equal amounts. In LEDCs it tends to be the young males.
- **Marital Status** In advanced countries most migrants have in the past been single. Nowadays family migrations are more common.
- **Occupational Groups** Professional migrants tend to predominate. Occupational migrations tend to be selective in terms of race, nationality and education.

Causes and characteristics

The conditions that cause migration can involve both 'push' (usually at the place of origin) and 'pull' (usually at the destination) factors.

Push factors:
- lack of wedlock
- increased mechanisation in agriculture
- low wages at origin
- highly political, racial or oppressive governments at origin
- natural disasters
- forced migration.

Pull factors:
- marriage
- employment offers
- the migrant has special skills
- retirement.

Models of migration

A series of models study **migration typologies**. They look at the distance travelled by a migrant, internal and international movement, the permanence of migration, the causes of migration and its selectivity.

Peterson's Typology of Migration

Identifies five classes of migration: primitive, forced, impelled, free, and mass movements, each with an activating force, and initiator. Each class of migration is further sub-divided into conservative migrants and innovating migrants.

Zelinsky's Mobility Transition Model

This is a five-phase model. In phase one there is just cyclic movement, in the second phase massive movement occurs, by phase four migration has levelled off. By phase five the only real movements are temporary inter-urban movements. Zelinsky's model closely mirrors the DTM. Zelinsky argued that migration is on the whole an orderly event and, though his theory is for the most part untested, it certainly seems to fit the patterns of migration seen in the developed world.

Lee's Laws of Migration

Outlines why groups choose to migrate. Summarises 'push and pull' ideas.

- There are factors linked to the destination of the migrants.
- There are factors associated with the origins of the migrants.
- Some intervening obstacles exist between origin and destination.
- Personal factors come into play.

Ravenstein's Laws of Migration

These were developed on the basis of migration for Great Britain between 1871 and 1881. He outlined eleven laws as follows:

- The majority of migrants go only a short distance.
- Migration proceeds step by step.
- Migrants going long distances generally choose a great centre of commerce or industry.
- Each current of migration produces a compensating counter-current.
- The natives of towns are less migratory than those of rural areas.
- Females are more migratory than males within the kingdom of their birth, but males more frequently venture beyond.
- Most migrants are adults; families rarely migrate out of their country of birth.
- Large towns grow more by migration than by natural increase.
- Migration increases in volume as industries and commerce develop and transport improves.
- The major direction of migration is from agricultural areas to centres of industry and commerce.
- The major causes of migration are economic.

Push and pull factors. Learn them!

Factors in the choice of destination:

- cost of moving
- presence of friends and relatives
- employment
- amenities
- features of the physical environment
- assistance and subsidies
- information available
- lack of alternative destinations.

Socio-economic factors influencing migration:

- technological change
- changes in economic practice and organisation
- propaganda
- regulations on migration
- healthcare and education
- population pressure
- climate and vegetation
- natural barriers
- size of countries.

Types of migration

Internal

Migrants move from areas of perceived deprivation to areas of perceived promise. Various criteria are assigned to internal movements in an attempt to classify them. Causes of movement tend to be either compulsory or voluntary, and they can also be temporary (like the cyclic movements of commuters in and out of a city) or permanent. Rural to urban is the dominant internal migration process, this depends upon a push component driving people away from the rural areas and a pull component, the so-called 'bright light' syndrome, attracting migrants to the cities.

Zimbabwe has experienced massive rural to urban drift since independence. Blacks had not been allowed to 'move' to the city when the country was ruled by the white minority. With independence, demand for jobs and housing very quickly outstripped supply. The capital, Harare, was rapidly overwhelmed. Knowing that Zimbabwe's rural to urban solutions depend upon a return to the land, the Government has started to plan for a rural future for its people.

International

Economic motives are most important in international migration. In general, a nation receives migrants from countries less developed than itself. Migrants tend to be either young and unskilled males or females, or highly skilled professionals. An alternative name for this type of movement is work or wage-related migration.

Gastarbeiter from Turkey migrated to Germany after the war. Germany needed to be rebuilt following the cessation of action against it in 1945 and needed to man its industry. Particularly through the 1960s, Germany had an almost open-door policy towards migrant workers, to ensure its economy continued to grow. By the eighties, 11% of the German workforce was foreign (27% of them from Turkey) and resident in greatest numbers in the southern industrial towns of Stuttgart, Frankfurt, Cologne and Munich. Germany has now closed its door to these foreign workers, and actively encourages them to go home, to ease pressure on its health, education and social services.

Over time the Turkish population has begun to cause problems for the Germans:

- They have little money for accommodation; they live in poor and overcrowded housing, which is unsightly; they send their spare money home.
- They refuse to fit into the culture of Germany, are unwilling to learn the language, etc.
- They create racial tension.
- They want to bring family members to Germany.
- The higher BRs of the *gastarbeiter* puts strain on Germany's resources, at a time when it is struggling to keep pace with the reforms that reunification has brought.

CASE STUDY

Refugees

A refugee is a person moved outside their own country because they are being persecuted because of their race, political or religious belief. If granted refugee status they become asylum seekers. There are in excess of 20 000 000 refugees in the world. The burden of housing these people being borne, on the whole, by the developing countries of the world. The Western approach is to tighten restrictions on migration. From Kosovo to Rwanda, from Afghanistan to Timor, from Ghana to Kurdistan there are many examples we geographers can draw on.

Who gains from migration?

Country of emigration

Gains	Losses
• less unemployment • money sent back helps raise living standards • migrants may return and bring back new skills to help development.	• most able and ambitious leave: they are best able to help development of their own country • family stress if split up for length of time • unemployment raised if returning migrants cannot find work.

Country of immigration

Gains	Losses
• jobs done cheaply • willing workforce to train • allows country to develop further • overcomes short-term labour shortage.	• money sent out of country • immigrants may not integrate: tensions build up with native people • problems of immigrants in times of high unemployment.

7.4 Population control

After studying this section you should understand:

- *why population policies are necessary*
- *what population policies hope to achieve*

Population policies

AQA A	U2	EDEXCEL A	U2
OCR A	U2	WJEC	U2
OCR B	U2	NICCEA	U2

When population projections are viewed they often spur governments into action. Policies influence growth, mortality, fertility, distribution and migration.

Anti-natalist policies

Methods include:

- providing contraception advice and devices/pills
- legalised abortion and late marriages
- economic and social measures to discourage large families, as in China.

Pro-natalist policies

These encourage population growth. Generally governments of pro-natalist states believe that their is strength in numbers and the economy will prosper.

Various countries have adopted such policies over the years:

- France with its 'La famille est prioritaire' scheme
- Romania under President Ceausescu
- Malaysia and Germany are two of many countries that offer maternity benefits and tax concessions for those families that enlarge.

A final thought... in the time it took you to read this section (*c.* 15 minutes) 700 children have been born in India alone.

Sample question and model answer

1

(a) (i) Describe the features of stage two in the demographic transition model.

[2]

> Improved medical care, sanitation and water supply means a rapid fall in death rates. Birth rates still remain high giving a rapid natural increase in population. Life expectancy increases.

Good answer. You could also comment on the difference between BR and DR.

(ii) With reference to **one** named country in stage two, describe the main features of its population pyramid.

[3]

> INDIA – The pyramid has a broad base indicating a high birth rate.

On the right track. Comment also that:
- *middle and old age are under represented*
- *increasing difference between BR and DR.*

(b) With reference to **one or more** named countries which discourage having children:

(i) outline the main characteristics and discuss the effectiveness of its population policy.

[8]

> CHINA – In the 1970's state family planning programmes were introduced: beginning with an advertising campaign for later marriages and fewer children. The next stage was to provide benefits for one-child families and penalties for multi-child families. The benefits included free education, priority housing and family benefits. The penalties included loss of these benefits and fines on the annual incomes. Couples also had to apply for permission to have a child and later in the 1980s abortions on second children became compulsory. These schemes stopped an increase in population of 55 million every 3 years in the 1960s and reduced birth rates from 40 per 1000 in 1968 to 17 per 1000 in 1980. Since the 1970s, family size has fallen from 5.8 to 2.4. However, these schemes have been more effective in urban areas due to better enforcement and education.

Good. Make it clear which country we are dealing with.

Strong on characteristics. Effectiveness less well covered.

Use of data is wise but not at the expense of answering the question.

Discussing effectiveness.

- effectiveness not really known because of falsification of records
- population has actually increased because 'baby boomers' of 'Great heap' are now bearing children
- less well supervised in the country, and as a result of this families move here
- people in rural areas frequently exceed the one child rules
- policy has encouraged female infanticide
- single children grow up spoilt
- single children are unable to support parents and grandparents.

(ii) State **two** ways in which you would expect the form of its population pyramid to change if the population policy was successful, so that the country progressed through the stages of the demographic transition model.

[2]

[AQA]

This question is to do with population pyramids. You should comment that:
- *the base is strengthened*
- *pyramid 'grows'*
- *more 'MEDC' shaped*
- *females living longer.*

This fails to focus on the question.

> 1 The country would reach stage 3 of the DTM with birth rates falling and death rate still falling.
>
> 2 Eventually stage 5 would be reached where birth rates fall below death rates giving a declining population.

Overall mark 8/15

To do better this student would need to focus on the questions and relate to the command words.

Practice examination questions

1 The diagrams below show population pyramids for an LEDC and an MEDC

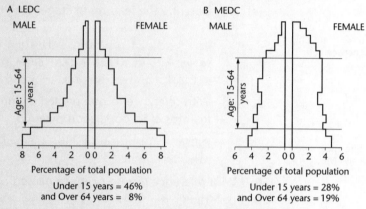

(a) Identify two major demographic differences between the pyramids. [2]

(b) Define what is meant by the term 'dependency ratio'. [2]

(c) (i) Suggest two future problems likely to occur in the LEDC as a consequence of the high percentage of its population under 15 years old. [2]

(ii) How might each of these problems be overcome? [2]

(d) The diagram below shows a population structure greatly influenced by migration. Some towns and regions within countries of the more economically developed world have this type of population structure.

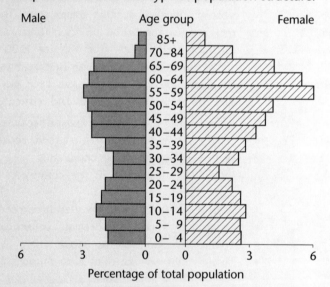

(i) Suggest and briefly explain what kind of town or region this could be if the dominant type of migration was: [4]

in-migration;

out-migration

(ii) Discuss some of the costs and benefits of an ageing population for a country. [8]

Practice examination questions *(continued)*

2 Study the diagram below, which is a model of international migration.

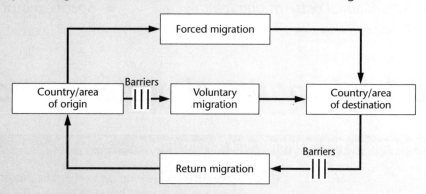

(a) (i) Define the term 'forced migration'. [2]

(ii) How and why might 'voluntary' and 'forced' migrations differ in terms of their:

population characteristics [3]

country of destination? [3]

(b) How might governments:

(i) Encourage 'voluntary migration' [3]

(ii) Discourage international migration? [3]

(c) With reference to examples, describe how physical factors might influence the volume of migration. [6]

[EDEXCEL]

Chapter 8
Worldwide industrial change

The following topics are covered in this chapter:

- *International patterns and trends*
- *Specific industrial examples*
- *Economic theories*

8.1 International patterns and trends

After studying this section you should understand:

LEARNING SUMMARY

- *how manufacturing can be defined by the industrial process carried out*
- *the recent trends in manufacturing at a global scale*

Classification

AQA A	U2	EDEXCEL A	A2
AQA B	A2	EDEXCEL B	A2
OCR A	A2	WJEC	A2
OCR B	U2	NICCEA	A2

Still a popular topic at AS!

If you asked anyone 50 years ago what the word 'industry' meant they might have replied 'mining and manufacturing'. Ask that same group of people today and their response would be vastly different. It is now one of many economic activities which include transport, communications, services, tourism and mining and manufacturing. Industry is work performed for an economic gain. To make the study of economic activity a little easier, a traditional classification or grouping of activities is often used and referred to, as below.

The biggest service industry is tourism.

> **Primary activity**
>
> Involves extraction or collection or the early processing of resources from activities such as quarrying, mining, forestry and farming. The activity is resource based and is the basis of secondary industry.
>
> **Secondary activity**
>
> Manufacturing, processing and assembly are the main activities; they change or transform the products of primary activity. The value of the raw material is increased by secondary activity, and locations do vary. Heavy industry adds bulk. Light industry is smaller scale, making TVs, etc.
>
> **Tertiary activity**
>
> Provides a service to customers. Such activities are market orientated. This area of work includes transport, shopping, teaching, office, banks, doctors.
>
> **Quaternary activity**
>
> The main service they provide is information and expertise. With the advent of computers, internet and fax, these industries are mobile and can locate just about anywhere. Many of these industries locate in universities, research and development centres. Marketing, advertising and research and development are other important areas.
>
> **Quinary activity**
>
> Includes education, Government, health and research.

KEY POINT

The diagram on page 127, based on **Fourastie's Model**, relates the approximate sizes of sectors today, based on their levels of economic development. It also indicates the approximate time-frame of industrialisation, de-industrialisation and re-industrialisation.

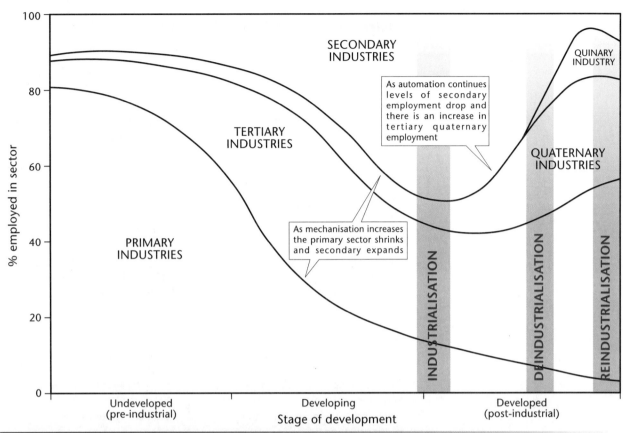

Trends and change

AQA A	U2	EDEXCEL A	A2
AQA B	A2	EDEXCEL B	A2
OCR A	A2	WJEC	A2
OCR B	U2	NICCEA	A2

Secondary or manufacturing industry has formed the cornerstone for most MEDC economies; leading countries through phases of economic development and on to a good deal of prosperity. However, the requirements of industry are changing, no longer do they need to be located at, or near, their raw material. They need locations that are spacious, with a rich labour market and that are accessible. This has led to fundamental changes in location to occur; the decline of 'regional' industry (e.g. the Ruhr, Germany and the UK West Midlands) and inner city and CBD sites. Popular now are the greenfield sites, city periphery sites. This fundamental change in location will continue apace as de-industrialisation is completed and rationalisation, restructuring, globalisation and re-industrialisation gathers pace!

The new industrial core of Europe

Manufacturing has undoubtedly declined in Europe – the service sector is now the largest employer and revenue producer. A variety of measures can be used to establish the new European core.

The highest GDPs are found from SE England through the Netherlands, Belgium and the western part of Germany. The poorest areas are the peripheral areas of Spain, Ireland, Portugal, Southern Italy and Greece. Unemployment shows a similar pattern to GDP. The term 'hot banana' has been coined for the core area; based on the joint shape of the countries involved. The area covers only 11% of Europe, but accounts for over 40% of output.

> The importance of the core and peripheral areas needs to be clearly understood at AS.

Other key factors include the effective and efficient transport network contained in this area, the skill of the labour force, and the system of tax and financial assistance the area accepts.

The biggest growth regions are the Paris, London and Amsterdam triangle; the Ruhr basin; various islands of innovation (e.g. Toulouse) and those areas that have attracted finance (e.g. Valencia, Ireland and Scotland).

Service society v industrial society

Service society	Industrial society
Differentiated production	Mass production
Short series of production	Long series of production
Flexibility and complexity	Standardisation
Economies of scope	Economies of scale
Knowledge and creativity most important factors of production	Capital most important factor of production
Development, planning, management and marketing most important functions	Goods production most important function
Quality maximisation	Cost minimisation
Product competition (quality, service, adaptation to customer needs)	Price competition
Turbulent and segmented markets	Stable and homogeneous market
International markets	National markets
Individual consumption	Mass consumption
Automation	Monotonous routine work
High and diversified qualifications	Standard qualifications
Job enrichment	Job specialisation
Flexible employment	Standardised labour market
Flat hierarchies, network organisation	Hierarchical organisation
Subcontracting, externalisation	Vertical integration
Small and medium-sized enterprise, divisionalisation	Large corporations
International cooperation, local and regional self-government	Nation states
Division of labour among firms	Division of labour among persons

De-industrialisation

This generally relates to the decline in jobs in the manufacturing sector of the employment structure. There are usually good reasons and cause, though it can be due to inefficient and inadequate production methods or infrastructure. A variety of such reasons are offered below:

- decline in the smokestack industries
- depletion of raw materials
- the cost of raw materials
- new technology and automation
- rationalisation
- a lack, or removal of subsidies
- the rise of the service sector in terms of employment
- competition from cheaper imports
- lack of investment
- diversification and mergers of larger companies
- manufactured exports don't match imports
- effect of over-zealous trade unions.

> Easily and increasingly examined at AS Level.

De-industrialisation:

- is coincidental with periods of depression
- may be massive and sudden and may disrupt the local balance of payments.

On the whole, de-industrialisation has been concentrated in the 19th-century industrial cities of Europe and the USA, and though it has meant the shedding of literally millions of manufacturing workers, what has been left in terms of manufacturing industry is much more competitive.

Industrial change in Corby

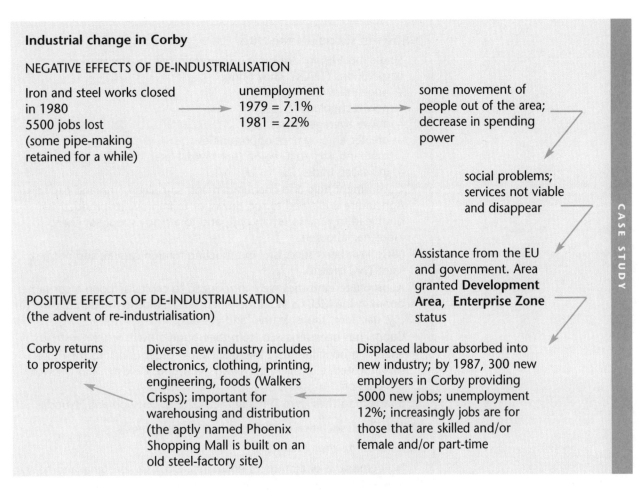

NEGATIVE EFFECTS OF DE-INDUSTRIALISATION

Iron and steel works closed in 1980
5500 jobs lost
(some pipe-making retained for a while)

→ unemployment
1979 = 7.1%
1981 = 22%

→ some movement of people out of the area; decrease in spending power

social problems; services not viable and disappear

Assistance from the EU and government. Area granted **Development Area**, **Enterprise Zone** status

POSITIVE EFFECTS OF DE-INDUSTRIALISATION
(the advent of re-industrialisation)

Corby returns to prosperity

← Diverse new industry includes electronics, clothing, printing, engineering, foods (Walkers Crisps); important for warehousing and distribution (the aptly named Phoenix Shopping Mall is built on an old steel-factory site)

← Displaced labour absorbed into new industry; by 1987, 300 new employers in Corby providing 5000 new jobs; unemployment 12%; increasingly jobs are for those that are skilled and/or female and/or part-time

De-industrialisation has on the whole been industry-specific, and invariably region-specific, e.g. Corby and area; Ravenscraig in Scotland; Durham and Tyneside, and so on. Most of these areas, thanks to Government and EU aid, are now experiencing the 'Post Industrial Age'.

Globalisation

The rise of the NICs (Newly Industrialised Countries)

During the 1980s, manufacturing began to shift globally to the NICs (and more recently to the RICs – **Rapidly Industrialising Countries**) of Asia and Latin America. This was due to a combination of de-industrialisation in the MEDCs and the successful combination of factors associated with the outstanding growth of the NICs, as shown on page 130.

The importance of high technology – the new global industry

Much of today's industry and economic activity is global in nature; or is multi-continental, with a range of activities taking place in many countries. The predominant feature of high-tech industry is its globalisation in the last 25 years.

This vital industry has massive effects and implications for the MEDCs of the world, its effect on our way of life is without question and its contribution to our wealth is paramount. High technology industries (e.g. ICT, telecommunications, consumer electronics, computer software, pharmaceuticals, etc.) all demand an innovative and committed workforce. The manufactured product is sophisticated and contains, or is the result of, the latest technology.

Globalisation is brought about by the TNCs who search the globe for the cheapest labour, the largest markets and lowest costs. Very roughly, one would expect microelectronics to dominate in the USA, Europe and Japan, and consumer production to dominate in the Asian NICs, though this situation is completely fluid and changes continually.

Causes of success in the NICs

- Single most important force is the growth of the transnational corporations (TNCs). They bring:
  - huge inflows of capital
  - new technology
  - above average wages
  - greater employment opportunities
  - more industry to a region (see Myrdal later in the chapter)
  - increased trade.
- Government help and encouragement, so-called command capitalism. Agriculture is allowed to run down, or to improve its productivity: both lead to rural to urban drift, and to a ready supply of cheap industrial labourers.
- NICs have been successful in attracting foreign capital, and not just from TNC growth.
- Appropriate industries were introduced to capitalise upon a consumer boom in the MEDCs at a time when MEDC industry was in decline (in toy, garment, shoes, leather and plastics).
- Capital has been amassed from their huge growth for reinvestment.
- Shifting production from low-quality low-order goods to the more sophisticated, e.g. microchips (e.g. Samsung, Goldstar and Daewoo) has paid off.
- Establishing their own TNCs in MEDCs, e.g. Daewoo and Hyundai.

Problems associated with rapid industrial growth

To do with the environment:

- high emissions of a range of pollutants, e.g. CO_2, bacterial and heavy metals
- deforestation.

To do with the economy:

- the growth of industry has left the infrastructure behind, e.g. intermittent electricity supply and poor road and distribution networks internally in many NICs
- manufacturing drains newly-won cash; there is a move, therefore, away from costly heavy industry
- TNCs take vast profits out of the host country.

To do with the people:

- workers are exploited; immigrant and child labour is common and maintains many 'economies'
- Western 'values' are abhorred in many of the Muslim NICs.

Positive spin-offs from NIC's growth

- foreign currency has allowed education, health, infrastructure, public health (sanitation, etc.) to be provided for and managed properly
- the number below the poverty line has fallen
- industry has developed; increasingly international markets are exploited and developed; not with low-order cheap goods, but with high technology/state-of-the-art products.

The future

- Asian NICs are likely to try to enter Chinese markets
- NICs increasingly widen their industrial base beyond that originally set up; loosen ties with MEDCs and their financial investment
- beware... the bubble bursting, e.g. South Korea's economic collapse in 1997, the run on the Yen in 1998 and 1999, and the knock-on effect on imports from Australia and the USA
- import control (i.e. GATT/WTO) is likely to impact on the MEDCs.

All extremely popular at AS Level.

KEY POINT

Governments across the globe also believe the high technology sector to be rather too important to leave alone. It is a massive revenue developer! They have contributed many millions of $, Yen and £s to ensuring its dominance in a number of countries.

Re-industrialisation – tertiarisation

The key area of industrial growth linked to re-industrialisation is tertiarisation – the growth of the service sector.

Service industries employ upward of 18 000 000 in the UK alone. The industry affects every part of the economy and de-industrialised economy, changing and adapting as different phases of the industrial process develop.

In the UK there is a high degree of inequality in terms of the amounts of service activity across the UK, because of:

- London's dominance
- the pre-dominance of the SE
- decentralisation.

Two different categories of activity are observed, as shown below:

Tertiarisation is so important in the UK/MEDC's economies that it frequently appears in AS questions.

Consumer services, which deal with the general public

Patterns of consumer provision have to match the locations and patterns of population. There are for instance fewer 'services' in rural West Norfolk than there are in and around Norwich. With regard to the location of such services, four important factors pervade:

- Services have to be extremely accessible, in terms of transport, but also telephonic links, fax and web connections need to be of the highest quality. The advent of advanced ICT has been crucial to the recent relocation/location of service industries.
- Invariably the proximity of the labour force can be a locating factor; they have to be close and there needs to be plenty of them! Service industries are major employers; many offer services 24 hours a day 365 days a year!
- Huge capital start-up costs mean that out-of-town locations are preferred nowadays: these sites are cheaper.
- Being footloose ensures that consumer services can be fairly flexible in terms of their approach to location and potential market.

On the whole, consumer services still tend to 'hug' centres of population.

Producer services, provide for other services

We are concerned here with management, public relations, computer services, etc. In terms of location much of business today is face-to-face, by way of discussion, or by phone, fax or video-conferencing. The SE of England in particular has seen a massive growth in this sector, being close to the capital's decision-makers and the research centres at the eastern end of the Thames valley.

As expected, producer services locate near their customers' industry.

In summary:

- It is feasible that a low capital start-up can occur with this industry.
- It is labour intensive, with the product being wholly dependent upon the standards applied by the staff and workers.
- Most jobs tend to be white-collared (i.e. non-manual).
- In a sense the user or purchaser of the service is an integral part of the production process.

Retail changes and developments (see also the settlement section)

Changes tend to mirror the movements of the population
i.e. urban to sub-urban to re-urbanisation.

Traditional location	Present location	Future strategy
• High-order goods providers, in the CBD or on the 'high street', probably family run. • Low-order goods providers, in shopping parades and 'local neighbourhood stores'.	• Corporate firms dominate, superstores are common, usually on brownfield sites, though some greenfield developments occur. • Retail parks, clusters of DIY, DIY, electrical and furniture stores.	• Shopping villages and outlet centres (selling labelled and over-run goods). • Consumer targeting (Airport shops, etc.). • Further development of Sunday trading and teleshopping. • Affected by government policy, the return of the 'centre' – the Mall.

Benefits include:

'Change' perceived as advantageous for customers by the companies involved.

Fall in the total numbers of shops, but there is an increase in floor space.

Change is explained by:
• small households, making small shops uneconomic
• increased affluence towards the suburbs
• car ownership and accessibility
• costly city-centre locations
• space for future development.

Retailing contributes about ½ of the 'service' GDP; the rest comes from banking and finance, the producer services.

Banking

In the producer service industry, banking in particular has gone through significant changes (mostly branch closures) due to the advent of new technology / tele-banking / increased use of ATMs / the advent of the credit card / banking via the internet.

Competition from new providers, e.g. Virgin Direct, Sainsbury's, Tesco and Direct Line Insurance all have a banking 'wing'. The move from building society to banking status (e.g. Abbey National) and recent mergers (Lloyds and TSB and the Bank of Scotland with NatWest) have also affected the industry. Many clerical tasks are now completed in less costly peripheral city sites.

CASE STUDY

CASE STUDY

8.2 Economic theories

After studying this section you should be able to:

- *understand how industrial theory has changed over the years*

Economic and industrial location models

AQA A	U2	EDEXCEL A	A2
AQA B	A2	EDEXCEL B	A2
OCR A	A2	WJEC	A2
OCR B	U2	NICCEA	A2

Theories of economic growth

Cumulative causation

The concept was developed by Swedish economist Gunnar Myrdal to explain the process that increases inequalities between regions. His model is displayed as a systems diagram. In Myrdal's original model, economic growth starts with new manufacturing industry. Of late it has been adapted to suit the new siting of service industry in the urban periphery.

The initial advantages of the new periphery site are likely to be strong infrastructural links and strong social acceptance of change. The expanding service industry attracts employment, and wealth begins to grow.

The increasing scale of activity on the new site has a multiplier effect, agglomeration occurs, and the site becomes increasingly successful. In the peripheral regions backwash causes a degree of downward spiral until spread effects reduce this imbalance.

The model has many merits. It seeks to explain the growth of industry as a series of linked relationships, on an undefined timescale. It works best at a national or international level but can be applied at local levels too, e.g. in Norwich Virgin Direct's Telesales Operation on the Norwich/Whiting Road Business Park has caused a number of cumulative effects but Norwich Union's operation in the city centre ensures the core continues to be vibrant.

Myrdal's model of cumulative causation

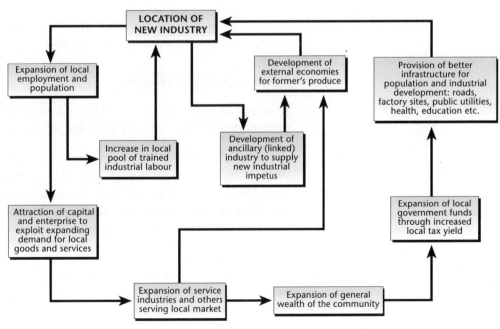

133

Rostow's model of economic growth

The economist W.W. Rostow devised a model that attempted to explain the development of economies in five linear stages or sequences. Rostow's graph plots time against expanding wealth.

Merits

- A useful starting point for understanding development.
- It can be used in conjunction with cumulative causation.

Limitations

- It is analogy based.
- It is based on Europe, USA and Japan.
- Growth in the economy can occur without development.

The Rostow model of economic growth

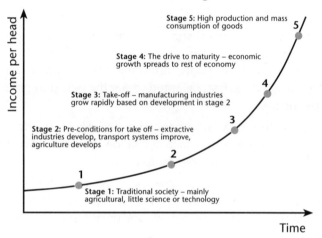

The product cycle as a four-phase model

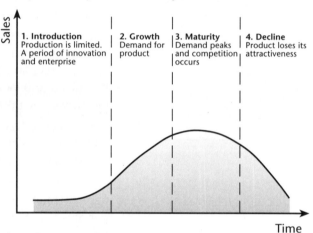

The product cycle

Based on work by R. Vernon, this cycle is used to study industrial location in relation to regional development.

Advantages

- It is simple.
- Everyday objects can be exemplified.

Disadvantages

- Assumes the product is created in its final form and that it doesn't evolve.
- Improvements to a product are ignored.

Early location theory

A variety of explanations and themes have been put forward in an attempt to explain industrial location. Chronologically they are:

Alfred Weber (1909)

Weber saw manufacturing as a series of least-cost locations or points. He considered the weight of raw materials, the weight loss or gain during manufacturing and transport costs. It was therefore a raw material/market oriented theory. He recognised the importance of agglomerations.

Harold Hotelling (1929)

Hotelling saw the manufacturing process as sites that would provide the largest revenue return. He recognised there would be competition, decision making and that producers locate at a market centre.

August Losch (1957)

The emphasis here was on demand, and that with distance demand would drop with increasing transport costs.

These three theories were based on assumptions about the physical landscape, transport costs and that man was a rational decision-maker!

Allan Pred (1967)

Worked not with locational decision-making but with an emphasis on behaviour. He recognised that businesses vary in the information that is available, how they use it and the importance of non-economic decisions. The size of companies is also important in Pred's theory. His conclusion was that man looks for a variety of micro-, meso- and macro-scale conditions before deciding on an industrial location.

> Be aware that industrial location has been studied for some 100 years and continues to be investigated.

David Smith (1971)

Smith's model defines the area where total revenue exceeds total costs. Industry will locate within the area in which profit can be made and not beyond the area where losses are made.

More recent location theory

'Smokestack is replaced by Sunrise Industry'.

Doreen Massey (1989)

Layers of investment theory, where the idea of inertia and missed labour opportunities is important to eventual location, e.g. the use of female labour.

Dunford (1989)

Sunrise industry locations are flexible in terms of material used, production, labour and transport costs.

Castree (1992)

Suggests that the footloose nature of industry means Weber's ideas are now redundant, with the exception of agglomeration.

Industrial location has evolved from economic factors, through behavioural to the geography of organisation. The models are still useful, though in a highly complex industrial world some simplification is essential.

8.3 Specific industrial examples

After studying this section you should be able to:

- *understand the state of primary and secondary industry in the UK today*

LEARNING SUMMARY

Primary and secondary industry in the UK

AQA A	U2	EDEXCEL A	A2
AQA B	A2	EDEXCEL B	A2
OCR A	A2	WJEC	A2
OCR B	U2	NICCEA	A2

The state of the car industry

The private car is big business in terms of employment and sales in the UK. However, our car-based economy is threatened by:

- changes in traditional production costs and markets

- environmental costs such as waste recycling, pollution and the use of land for parking and roads.

The market for cars grew steadily in the early 1980s. But since 1988 the British motor industry has changed rapidly with up to 50 000 losing their jobs in the UK industry alone and nearly one million car sales disappearing almost overnight despite increased productivity in the UK. Recession, and our departure from the ERM, can account for the bulk of these changes.

By 1993, Ford announced that it was losing £1 000 000 per day and, with recession biting hard in Europe, UK car sales to the continent all but stopped.

Faced with increased costs and resulting redundancies the industry had to make changes.

1 Mergers occurred: Rover with BMW; VW with Audi, Seat and Skoda. These make possible massive economies of scale, and make the companies extremely competitive.

2 JIT (Just-in-time production). This system, devised by the Japanese, has had a remarkable effect on car production, reducing costs and overheads. Under this system, computers with direct connections between factories can be programmed with the full production plans and can specify exact requirements at precise times. Delivery systems span the world. This system negates the need to tie up vast amounts of capital in stocks at the factory. Nissan UK estimate they have cut 2.4 million miles off the delivery mileage for the Micra through this system, with savings being passed on to the motorist.

The British car industry looks set for a future in research and development rather than manufacturing. (In May 2000 Ford announced the closure of its Dagenham Fiesta-producing plant.)

> Primary and secondary industry is not so widely covered at examinable AS Level. But a knowledge of what has gone before is vital!

The state of coal mining

Industrial development was based on coal in the UK, but gradually coal's predominance has been challenged by oil, despite an aggressive Europe eager to sell its supply of cheap coal to many of the UK's traditional customers. As in Europe, the UK coal industry has grown and expanded rapidly, collapsed and wrecked lives and economies.

One hundred years of mining

- **1900** Continued expansion and the peak of coal production to a growing and energy-hungry heavy industry.
- **Since 1945** Residential and industrial use has collapsed. Replaced by power stations. Improvements in technology and mechanisation have closed high-cost, low-production pits.
- **Since 1960** Cheap oil purchases from the Middle East reduced demand further.
- **1973** Energy crises ensured the Government's determination to develop other cheaper sources of power.
- **1974** Last-ditch attempt to 'rescue coal', with flagship projects at Selby and the Vale of Belvoir.
- **Early 1980s** Imports flood onto the UK market from Europe and further afield.
- **1984 to 1985** Miners strike, the death knell for coal.
- **1990s** Concern for the environment leads to a reassessment of coal's future use in the UK.
- **1992 to 1993** 31 pits closed, 30 000 lost their jobs.
- **1994** 18 000 miners now employed. When the industry was nationalised in 1947 there were 750 000 miners!

Sample question and model answer

1

The diagram below gives some information about South West England in general and Cornwall in particular.

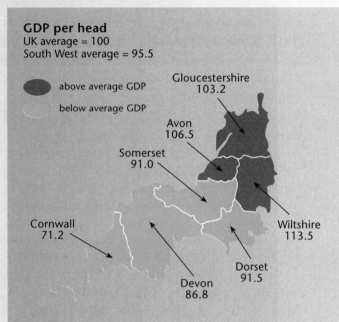

GDP per head
UK average = 100
South West average = 95.5

above average GDP

below average GDP

Gloucestershire
103.2

Avon
106.5

Somerset
91.0

Cornwall
71.2

Wiltshire
113.5

Dorset
91.5

Devon
86.8

In Cornwall, where the decline of the mainstay fishing and tin mining industries has led to more deep-seated structural problems, the task of attracting outside investment has been handled for more than a decade by Devon and Cornwall Development International, which has successfully targeted companies in the healthcare, medical products and electronics industries.

Source: adapted from The Guardian, 22 August 1996

(i) Describe the pattern and the size of variations in wealth within the counties of South West England. **[3]**

Gloucestershire, Wiltshire and Avon are the three wealthiest counties with a GDP above that of the national average. They are grouped together with the two wealthiest nearest to those areas which fall below UK average. There is a general decline from North East to South West. The three wealthiest counties in the South West equal approximately 1/2 of the area of the poorest counties in the South West. Cornwall has a GDP of 71.2 (28.8 below UK average) compared to Wiltshire which has a GDP of 13.5, above the UK average.

Author's note: For a third mark a comment on range and average values is needed.

There is a general decline from East to West.

Compares relative positions of counties behind Wiltshire.

(ii) Why would it make sense for areas such as Cornwall to reduce their dependence upon primary sector industries? **[3]**

By reducing their dependency on primary industries such as forestry, fishing and mining the workforce would be available for newer high tech industries. The area would be more attractive to future developers who may be inclined to move in.
Tourism plays a major part in the survival of Cornwall and introducing 2° and 3° industries which give a higher financial return to the area is the way forward.

Author's note: For third mark more needed on general problems of primary sector.

Recognition that fishing and mining are primary industries.

Idea of higher return from a different sector.

Sample question and model answer (continued)

(iii) Why is it difficult to classify new sources of employment, such as those which have moved into Cornwall, into particular employment sectors? [2]

Not specific enough: 'idea' that healthcare could be secondary, tertiary or quaternary: And that it doesn't fit standard industrial classification.

Industries which have moved into Cornwall tend to be geared towards providing services. These services can range from a new hospital to transport and entertainment or policing. Due to tourism, these industries are not in great demand all year round.

(iv) For one named area in which manufacturing industry is important, give reasons for the decrease in employment opportunities within the manufacturing sector. [7]

Decline in numbers employed, related to declines in manufacturing industry and mechanisation.

There has been a decrease in employment opportunities within the car manufacturing industry in the Midlands for several reasons. The first is the gradual withdrawal of foreign investment. This has lead to some companies becoming bankrupt and making their employees redundant. The unemployed who are not already trained and skilled in a specific area find it particularly difficult to find jobs in an industry which is rapidly becoming dominated by more efficient machines. The jobs which are available require highly skilled personnel, of which there are too few ready trained.

Competition.

'Ideas' to do with rationalisation.

Recently, there has been a decrease in demand for new cars, resulting in overproduction and heavy losses for the company involved. In some cases, a large number of employees have been forced to take redundancy. It is possible to find employment elsewhere in other manufacturing industries which use similar techniques and production methods, for example metal casting or the production of sheet metals. However, these industries often go 'hand in hand' with each other, so any problems arising within and affecting one, will inevitably affect the other, thus a decrease in production of cars leads to a decrease in the workforce needed, which in turn may lead to a decrease in demand for sheet metal or cast metal engine blocks and so eventually, a decrease of labourforce in the parallel industries.

Change in demand.

AEB/AQA

Practice examination questions

Section A

1 (a) Define the terms (i) *de-industrialisation* and (ii) *branch plant economy*. [4]

(b) For **one** named region where de-industrialisation is occurring, outline the effects of the process on the regional economy. [6]

(c) Explain why a branch plant economy does not guarantee continued economic development. [5]

[AQA]

Section B

2 (a) With reference to selected examples, account for the reduced importance of primary economic activities in MEDCs. [10]

(b) How true is it to say that primary economic activities have far less environmental impact than secondary activities? [5]

Practice examination answers

Chapter 1 Water on the land

Section A

1 (a) You could talk about the relationship of lag time to precipitation events, peaks of rainfall, shape of the peak, base flow separation.

(b) Possible factors include: vegetation, soil state (i.e. is it saturated or dry?), type and intensity of precipitation, slop, nature of the vegetation, climate, etc.

(c) Because rate of throughflow is constant through the adjacent rock.

(d) (i) Shape of the basin, gradient
(ii) Shape of the basin:
circular basin – delivers water consistently, and is released from the system slowly.
bell shaped basin – delivers water rapidly, leading to flashy/peaky shape to the hydrograph.
Gradient:
steep – flashy/steep shaped hydrograph profile.
shallow – gentle = bell-shaped hydrograph profile.

(e) Hourglass-shape basins have two peaks.
Areas with variable geology may have one peak from runoff and one from groundwater release.

(f) More water = more energy = more erosion and more transport.
The shape of the channel is deepened and the bed debris is transported away.

(g) Hydrographs can be drawn of any part of the basin. They indicate how interception, infiltration and consequent runoff is affected. This aids flood control in a river basin as hydrologists can plan for extreme events related to the flashiness of the hydrological profile.

Section B

2 Because both play an important part in determining size and frequency of flooding.

Factors include:

Physical
Rock type – determines permeability
Relief – affects run-off rates
Precipitation and climate – severity and frequency important
Snow melt – releases large volumes of water at one time
Confluence effects – surges of water coinciding in one place.

Human
Intensity of urbanisation on the floodplain – built surfaces don't absorb water
Agricultural land use – affects infiltration
Afforestation and deforestation – affect rates of interception
Interference and manipulation of the river – affects speed of discharge, levées may prevent flooding.

Examples include, and of course there are others!!
In the UK = Ouse in Northamptonshire and Yorkshire, Severn in Shropshire.
In the USA = Mississippi, 1993 flood.

Chapter 2 The challenge of the coast

Section A

1 (a) (i) The best answer would closely resemble the beach cross-section in the coastal chapter!

(ii) • Nature of beach sediment
• Storm recurrence
• Nature and extent of cliff decline
• Other acceptable labels.

(b) (i) • Direction of movement needs to be marked on
• Recession of the cliff line needs to be marked on
• Accumulations against the piers appropriate to drift position must be marked on

(ii) Explanation primarily involves comment on the restrictions of movement caused by the piers. Look for a well thought through explanation.

(c) You should have discussed the problems of interfering in a complex natural system and the occurrence of unforeseen impacts down coast, e.g. increased or decreased erosion and deposition. Examples are a must. Quality answers will be convincing in terms of policy towards conservation and protection.

Section B

2 For the answer, see the sea level change sample question earlier in the coastal chapter.

Chapter 3 The challenge of the atmosphere

Section A

1 (a) • named Hadley Cell rising and falling between 0 and 30°N or S
• circular pattern of rising and descending air and indicated for the other cells too
• cells could be named.

(b) Heat source:
• influence on the jet stream
• thinning of tropopause towards the poles
• differential heating and cooling of the air.

(c) Air rising:
• heating of surface

- high insolation and uplift
- air masses meeting etc....

Cloud type:
- at 0° cumulus/cumulo-nimbus
- further north stratus
- different heights of clouds.

Precipitation:
- short torrential rainfall
- some slight rain.

(d) Any three from
- increase in violent and unpredictable weather – flooding and hurricane intensification
- when warm water replaces cold there is decimation of fishing grounds
- drought in normally lushly vegetated areas
- agricultural production plummets
or other suitable answers.

(e) Effects on agriculture, fishing, construction, transport, power, business and retail, leisure and sport, health.
Could be + or – effects.

Section B

2 (a) General 'impact' points:
- temperature increase affects precipitation
- soil moisture drops
- drought problems
- agricultural output drops, especially in poorer tropical countries
- increased cloud cover reflects insolation, shading and cooling the ground
- sea level increases, through ice melt and thermal expansion of the sea
- wetlands dry up
- ocean currents change
- jet streams change position
- extinction of natural flora and fauna, in various countries
- increase in tropical diseases, and so on... .

Consequences – 'economic and social':
- snow-cover reduction – tourism/skiing affected
- water demand and supply affected
- location of fish shoals change
- regional weather shifts
- more intense, less-predictable storms
- pests spread
- different crops grown.

(b) General points using the UK as the exemplar:
- temperate climate becomes more Mediterranean
- less-frequent heavy rainstorms replace frequent drizzle and frequent rain
- insect infestation increases
- sea level rise exacerbated in the SE of the UK, because of global warming and continued Pleistocene isostatic re-adjustment
- warmer S. UK would support citrus groves and cooler Scotland dairying
- coastal areas inundated, nuclear power stations under threat, coastal industry retreats inland and the coastal population migrates too
- salt marsh disappears
- coastal erosion increases.

Of course, the answer could focus on any country of any continent.

Chapter 4 Earth challenge

1 (a) The break down of rocks *in situ* (1 mark) by physical or chemical or organic activity.

(b) Any suitable process and how it works.

(c) More moisture is available and temperatures vary.

(d) The movement downslope of regolith and soil under gravity as influenced by water or ice.

(e) One of: solifluction, rock falls or soil creep.

(f) Influence is from seasonal freeze and thaw – heave and rockfalls occur.

(g) Two from:
- gradient of slope
- vegetation cover
- rock type and lithology
- human activity.

2 (a)

	(source)	(cause)	(effects)
Neuées Ardentes	Occur on destructive boundaries.	Sudden release of mixed hot rock, lava particles and ash buoyed by hot gases.	Fast-flowing, incandescent cloud. Destroys everything in its path, e.g. Soufriere Hills, Montserrat 1996.

	(source)	(cause)	(effects)
Air fall tephra	Magma	Small pieces of ash and tephra form when solid magma is broken by groundwater turning explosively into steam.	It is so fine it is carried huge distances. Causes breathing difficulties, building collapse. Agricultural losses, etc, e.g. Pinatubo 1991.

(b) • Agricultural potential; extrusive materials weather to fertile soils
- Tourist value
- Building materials
- Possible geothermal power
- Poor have nowhere else to live.

(c) Earthquakes are a good example.

(adjust)
They adjust behaviour to modify vulnerability; predict/warn/educate; prepare community to evacuate; modify landuse planning; set up aid insurance schemes; use seismographs.

(modify/cope)
They make structural changes, bracing buildings; foundation work; special structure design, etc.

Chapter 5 Soils and ecosystems

1 (a) A = Topsoil
B = Subsoil
F = Fermentation layer
H = Humus layer

(b) Drainage and aeration are affected by the soil structure.
Leaf fall and subsequent decomposition affect the depth of the O horizon.

(c) Gleyed soil - waterlogged and mottled.
Podzols - clay accumulation, is red. Hardpan in place.
Tropical soils - red/yellow in colour. Sesquioxide accumulation.

(d) pH is acid from fulvic acid. Released from vegetation. Rain is slightly acidic.
Characteristics of the A horizon can be altered by deep ploughing, adding lime or manure.

(e) On well-drained brown earths the planting of conifer plantations can initiate podzolisation. Over-cultivation also acidifies with time.

2 (a) X = leaf litter; Y = litter storage.

(b) Any two from calcium, magnesium, phosphorus and potassium.

(c) Biomass is the total amount of living organic material in the ecosystem.

(d) Nutrients are transferred from soil to biomass in solution – roots taking up soil water.

(e) Any three from:
- reduction in biomass = reduced humus
- reduction in humus leads to a breakdown of the clay–humus complex
- no clay–humus complex = breakdown of soil structure
- reduced vegetation = more runoff = more erosion.

Chapter 6 Settlement issues

Section A

1 (a) (i) We have to decide how our cities are to develop in the future because of the role they play in our lives.
- Because of growing dis-economies in the city
- Changing aspirations
- Technological change
- Development in cities costs, money has to be spent properly
- The cities should not become the reserve of just the rich in isolation
- Green avenues or 'hearts' have to be maintained
- There are social and environmental prices to pay if wrong decisions are made
- Should green field sites be avoided in the green-belt, and new towns be established instead?

(ii) Lead by government. There is little local input.
- Governments control the purse strings. There is a need to move into private financing of urban development schemes
- Government planners pinpoint/earmark land for development. This should be in the hands of the local planners
- Most planning is focused in and around urbanised areas. The 'spread' needs to be dispersed
- Planners build and provide housing for the middle-class owner-occupiers. Affordable housing needs to be built.

(b) (i) The need for green belt retention.
- Most problems are to do with conflict between development and conservation
- Green belt retention revolves around a need to stop urban sprawl, to stop town and cities merging, to provide recreation, to safeguard agricultural activity, preserve the historic character of towns and to assist urban regeneration.

(ii) Reasons why the green belt has to be used.
- Development is not then forced beyond the green belt
- So the citites are not so densely packed
- To avoide house and land price hikes near the green belt
- To restrict community and pollution
- Fewer roads need to be built
- Establishing country parks in the green belt ensures their preservation
- Green belts buffer the countryside from the pressures of population growth
- Green belts must accommodate shortages of city housing
- Land-use zoning is necessary to try to discourage the stranglehold of the green belt
- De-restriction of the green belt has now been written into the statute books to allow employment opportunities/residential areas and roads, etc. to be developed
- Many of the allowable green belt activities (rubbish dumps, intensive agricultural activity/stadia development) have left the green belt scarred and in need of development.

(c) (i) Include: families wanting a safer, more pleasant environment.
- Small settlements are preferred by an increasing number of people
- Retirees want quiet/calm/clean environments
- Small-scale entrepreneurial activity is increasing in the rural areas because labour is cheap and the rural environment is more appealing

(ii) Rural areas are shrinking in size.
- To prevent the spread of urban blight, noise and pollution
- To prevent the spread of road systems
- To prevent urban leap-frogging
- To prevent dilution of the rural idyll

Comments could be biased towards the positive (to prevent sprawl) or can work to promote sprawl of the city (see b's answers).

(d) Factors include:
- **Environmental**: discharge the need to drive, insulated housing, recycling sites, build utility efficient housing
- **Social**: rentable housing as well as owner occupied housing, secure housing, disabled provision, link jobs to home, build neighbourhood and community housing
- **Demographic**: build suitable housing in high-density population areas, build for the pensioner and single person occupiers.

Section A

2 (a) Definition of village:
- site explained
- range of acceptable physical factors covered
- growth by exploitation, explained (of soil, timber etc)

- local and global examples used
- may introduce a counter argument to do with planning, politics (set-a-side) and the social angle (key villages scheme). Balance of site v growth ideas necessary.

(b) Changes in:
- form – infill, linear to ribbon development, barn conversions.
- function – commuter / dormitory villages. Attraction of new functions, the baker swops premises with an antique shop for instance.
- sociological change – effect on house prices, 'townies v farmers'.
- environmental conditions – noise, pollution and loss of land.

Reasons:
- better access
- affluence
- push and pull factors
- retirees

You might also refer to LEDCs.

Chapter 7 The dynamics of population

1 (a) LEDC increasing BR but MEDC BR stable
LEDC patterns for male and female situation
MEDC – Female greater life expectancy.

(b) The proportion of young and old people to the economically active.

(c) (i) Increase in BR; unemployment
(ii) By family planning measures or limiting families to one child
By job creation schemes, public investment, increasing time spent in educational pursuits.

(d) (i) In-migration;
Holiday resort or retirement area (1 mark)
Refers to wealth/mobility.
Uses values from the graph (1 mark)
out-migration
Declining area/area where young have left (1 mark)
Uses values from the graph (1 mark)

(ii)
Costs	Benefits
Elderly are expensive for health service and state pensions, etc.	Private pension holders Have a lot of disposable income for holidays, home improvements.

2 (a) (i) Migration which is compelled either by government or other factors.
Exemplification will allow access to the extra mark.

(ii) Population characteristics:
Voluntary migrations will be young, male and single and the forced migrations might include

all people possibly excluding men. Men will seek employment opportunities as a result of the voluntary migration whereas in a forced migration all the people will have to move – the men might be killed as a result of the war, e.g. Kosovo.

Country of destination:
The voluntary migrations will mainly be to MEDCs or richer areas. Migrants will be often seeking work and these areas contain more employment opportunities. Forced migrations will often be to a neighbouring area.

(b) (i) Subsidised fares, e.g. £10 Poms to Australia.
Relax immigration requirements for key workers – doctors, scientists, teachers to NZ
Encourage free movement of personnel across frontiers – EU

(ii) Imposition of passport controls
Deport illegal immigrants
Put a property or qualification barrier to immigration – NZ
Work permits – 'Green Card'.

(c) You should discuss the physical conditions in the out-migration country, e.g. forced migration through drought or soil exhaustion.

You might also look at the physical conditions in the receiving country, e.g. the presence of a more attractive climate in California.

Add points that relate to voluntary and forced migration and discuss the friction of distance.

Chapter 8 Worldwide industrial change

Section A

1 (a) (i) De-industrialisation takes place when a countries industrial base shrinks.

(ii) Usually found in peripheral areas where a large proportion of employment in manufacturing is in extremely controlled branch plants.

(b) • E.g. West Yorkshire woollen Industry.
  • Effects
    = decline in manufacturing employment
    = contribution to GDP drops
    = unemployment
    = dereliction
    = service decline
    = housing stock deteriorates.

(c) Because, during recession or severe competition large firms implement new policies. Restructuring occurs and branch plants that are labour intensive rather than capital intensive are closed, as decisions are made elsewhere.
Branch plants are common in 'peripheral' Europe. The proper name for this closure is organisational disintegration.

All subsections are point marked, but look for sensible development.

Section B

2 (a) Define *primary* economic activity and for selected examples, discuss their contribution to GDP, numbers employed, compare with other industry types. A good answer might refer to theory, e.g. the Development Stage Model.

Possible references include:
• Exhaustion of resource – e.g. fish, iron ore in Britain

• Cheaper to import resource, e.g. UK – coal
• Decline of manufacturing and hence less need for primary resources – shift to tertiary/quaternary activities
• Some effect of substitutes – synthetic fibres, artificial sweeteners
• Problems of over-production in agriculture – EU

N.B. You could question the veracity of the statement, e.g. UK – oil/gas, Australian/Canadian – mining, etc.

(b) • This has an evaluation slant – your answer should show depth of understanding, logical arguments and balance.
• Impact of primary activities: destruction of habitations, waste disposal, pollution of water courses, soil erosion, desertification, over-fishing, etc. – relate these to appropriate activities.
• Impact of secondary activities: can indirectly be responsible for the above but also – loss of land, waste, pollution (of air, water, land); elaborate these, e.g. acid rain. Traffic generated is also a contributory factor.

N.B. Secondary activity tends to be concentrated spatially – effects are consequently exaggerated.

Internet interest sites

To keep up to date, browse the web occasionally, there are literally 100 000s of geographical link sites available. Below are just a few of the sites visited during the preparation of this book.

1 Water on the land

A journey down the LA river –
http://lal.k12.ca.us//aep/start/river/tour/index/html
Fact sheets – http://www.fema.floodf.html
General –
http://terassa.pnl.gov:2080/EESC/resourselist/hydrology.html

2 The challenge of the coast

MAFF – http://www.maff.gov.uk/
Atlantic Oceanographic Survey –
http//landsea@aoml.noaa.gov

3 The challenge of the atmosphere

Met office – http://www.meto.govt.uk/
National Hazard Centre –
http://www.nhc.noaa.gov/products.html
Greenhouse Effect –
http://www.dar.csiro.au/pub/info/greenhouse.html
El Niño – http://www.pmel.noaa.gov/toga-tao/el-nino/forecasts.html

4 Earth challenge

Landslides –
http://geohazards.cr.usgs.gov/neis/general/handouts/faq.html
South California Quake Centre –
http://www.scecdc.scec.org/

Volcano World – http://volcano.und.nodak.edu/
The world's most active volcanoes –
http://www.geo.mtu.edu/eos/
Earthquakes –
http://gldss7.cr.usgs.gov/neis/general/handouts/faq.html

5 Soils and ecosystems

By far the best site is that operated by Reading University, it is full of interactive information.
Rainforests – http://www.ran.org/ran/

6 Settlement issues

UK Government Index –
http://www.open.gov.uk/index/findex.html
Environment Agency – http://www.environment-agency.gov.uk/

7 The dynamics of population

US Census Information –
http://cedr.lbl.gov.cdnon(lookup!).
Population Concern's web page is of real use to the geography student. You must visit it.

8 Worldwide industrial change

Sheffield Cultural Industry Quarter –
http://syspace.co.uk/ciq/
Most industrial manufacturers have their own web pages that you can visit.

Most, if not all, of the above have interactive links on their web pages.

Index